Yabuki Yūji, 18th Sōke of Ono-ha Ittō-ryū and Executive Director, Reigakudō

Ittō-ryū was created in the Muromachi Period by Itō Ittōsai Kagehisa and then passed on to Ono Jirōemon Tadaaki.

After Tadaaki became the official kenjutsu instructor of Tokugawa Hideyoshi, the House of Ono assumed the responsibility for providing official instructors to successive generations of shoguns, cementing Ittō-ryū as the epitome of Japanese swordsmanship. Many schools branched off of Ittō-ryū, and the mainline was called Ono-ha Ittō-ryū to distinguish it from those other factions. After being transferred temporarily to Tsugaru Nobuhisa by Ono Tadakazu, the mainline school returned once again to the House of Ono. After that, it was passed down through both the Tsugaru and Ono families, and then further bequeathed to the House of Yamaga from Ono Tadayoshi. In the beginning of the Taishō Period, the mainline was conveyed to Sasamori Junzō. The above, to include new techniques that were added by Tadatsune and Tadao, laid the foundation for our current instructional methods, training approaches, and scrolls. Sasamori Junzō traced these various transmission lines back to their origins, and using his research as a guide, ensured that the techniques and theories of all of them were reintegrated back into the mainline tradition, forming the Ittō-ryū of today. After that, he passed it to Sasamori Takemi.

The main material that Sasamori Junzō gathered was published in his book, *Secrets of Ittō-ryū*, but now, with the cooperation of Kokugakuin University and the Japanese Sword Museum, we were able to compile treasured documents he inherited that spanned from the Edo to the Taishō Periods and publish them in this book as the *Ono-ha Ittō-ryū Patriarch's Repository: Sword and Document Compendium*, a collection of photographs and commentary on these cultural artifacts.

While examining this sword collection, we discovered a hitherto unknown short sword signed by the swordsmith Kaneie that dates from the Muromachi Period. It was a significant find that evokes the lore of Itō Ittōsai's time.

We are also honored to present new material, such as letters from Saimura Gorō, Ogawa Kinnosuke, Naitō Takaharu, and Takano Sasaburō that show Ittō-ryū's close links to modern kendo.

It is my hope that many people will gain a deeper understanding of Ono-ha Ittō-ryū through this book.

References

Sasamori Junzō. *Ittō-ryū Gokui.* (Reprinted edition.) Tokyo: Taiiku to Sports Suppansha, 1986.

Negishi Yasumori. *Hasegawa Tsuyoshi Kōchū. Jinō* (Vols. 1-3). Tokyo: Iwanami Shoten, 1991.

Shoji Munemitsu. *Kendo Hyakunen.* (Revised edition.) Tokyo: Jiji Press, 1976.

Wataya Kiyoshi. *Nihon Kengo 100 Sen.* Tokyo: Akita Shoten, 1983.

Nihon Budō Gakkai Kendō Senmon Bunkakai. *Kendō wo Shiru Jiten.* Tokyo: Tokyodo Shuppan, 2009.

Contributors

Photography:	Onishi Hideaki (FLAG Co. Ltd.)
Design:	Mitsutake Masaki
Translator:	Mark Hague (Alkaid Research LLC)
Editor:	Lee Mamie
Technical Contributors:	Ishizaki Toru (Sōke of Shin Musō Hayashizaki-ryū Iai)
	Yabe Kentarō (Professor, Department of Literature, Kokugakuin University)

Author:	Yabuki Yūji, 18th Sōke, Ono-ha Ittō-ryū
	Reigakudō General Incorporated Association
	Daizawa 1-13-2, Setagaya-ku, Tokyo, 155-0032
	https://onohaittoryu.3.pro.tok2.com/index_en.html
Published by:	Alkaid Research LLC
Publishing date:	©2023. Originally published in Japan as *Ono-ha Ittō-ryū Sōke Shozō Tōken Bunshoshū* by Yabuki Yūji, 18th Sōke, Ono-ha Ittō-ryū, Reigakudō General Incorporated Association. All rights reserved.

ISBN: 978-4-911080-00-9 C0070 (2023 original edition)

ISBN: 979-8-9872421-4-8 (2025 English language edition)

Table of Contents

Lineage

- Ono Jirōemon Tadaaki, the founder of Ono-ha Ittō-ryū, inherited the mainline of Ittō-ryū through direct transmission from Itō Ittōsai Kegehisa, the progenitor of Ittō-ryū. To differentiate it from other branches and factions of Ittō-ryū, it would be called, Ono-ha, the Ono faction. Tadaaki became the official instructor of Shogun Tokugawa Hidetada.

- Ono Jirōemon Tadatsune was the third son of Tadaaki. He was originally known as Tadakatsu. After he learned the mainline from his father, he inherited the name Jirōemon. In addition to Tokugawa Iemitsu, Tadatsune also taught many other students.

- Ono Jirōemon Tadao was the fourth son of Tadaaki and became the adopted son of Tadatsune. Another theory is that he was one of his disciples. After learning the mainline from Tadatsune, he performed duties as the official kenjutsu instructor to Shoguns [Tokugawa] Ietsuna, Tsunayoshi, and Ienobu.

- As the 4th Lord of the Tsugaru Domain, Tsugaru Nobumasa learned Ittō-ryū from Tadao, mastered its secrets, and became famous for his mastery of the military and literary arts.

- Ono Jirōemon Tadakazu, who was initially known as Okabe Sukekurō, changed his name to Tadakazu after he was adopted by Tadao, learned Ittō-ryū from him, and carried on the lineage. Tadakazu, aside from teaching the Tokugawa Shoguns, also had many other students and conveyed the mainline of Ittō-ryū to Tsugaru Tosa-no-Kami Nobuhisa instead of his own son.

- Tsugaru Tosa-no-Kami Nobuhisa, the 5th Lord of Tsugaru, mastered the inner secrets of Ittō-ryū from Tadao, and by a decree from Tadakazu, inherited the school as the sole heir. With this decision, the direct transmission of Ittō-ryū temporarily passed from the hands of the House of Ono to the House of Tsugaru.

- Ono Jirōemon Tadahisa was expected to inherit the school after being taught by Tsugaru Tosa-no-Kami Nobuhisa but passed away unexpectedly at a young age.

- Ono Jirōemon Tadakata was still young when his father, Tadahisa, passed away so he wasn't able to learn Ittō-ryū from him. Lamenting the fact that Ittō-ryū may die out, Tsugaru Nobuhisa (retirement name Sakae Ō) came out of retirement in his old age and taught Tadakata as he got older, thus preserving the school by passing on the mainline to him.

- Ono Jirōemon Tadayoshi was the son of Tadakata. He learned from his father and taught many of his own students.

- Ono Jirōemon Tadataka was the son of Tadayoshi, learned from him, and succeeded him as the head of the mainline.

- Ono Jirōemon Tadasada was the son of Tadataka, learned from his father, and fully inherited the mainline from him.

- Ono Nario learned from his father and received the mainline from him. (Ono Nario was the last of the Ono family to teach Ittō-ryū.)

- Nakanishi Chūta Tanesada learned Ittō-ryū from Ono Jirōemon Tadao and Tadakazu. (Nakanishi-ha Ittō-ryū.)

- Yamaga Sokō was taught Ittō-ryū from the founder, Ono Jirōemon Tadaaki. (Yamaga Sokō taught military science, i.e. strategy, to the Tokugawa Shogunate and Houses of Tsugaru and Ono.)

- Members from the House of Yamaga became official kenjutsu instructors to the House of Tsugaru while maintaining contact with the Houses of Ono, Tsugaru, and Nakanishi.

- During the Taishō Period, Sasamori Junzō received full transmission of the mainline of Ittō-ryū from the Houses of Tsugaru and Yamaga. After combining what he learned from both into one curriculum, he became the 16th Sōke of Ono-ha Ittō-ryū.

- Sasamori Takemi learned Ittō-ryū from his father, Sasamori Junzō, and later became the sole heir to the school, succeeding him as the 17th Sōke.

- Yabuki Yūji learned from Sasamori Takemi, succeeding him to be the sole heir of the school and 18th Sōke of Ono-ha Ittō-ryū, a position he currently holds.

Lineage Chart

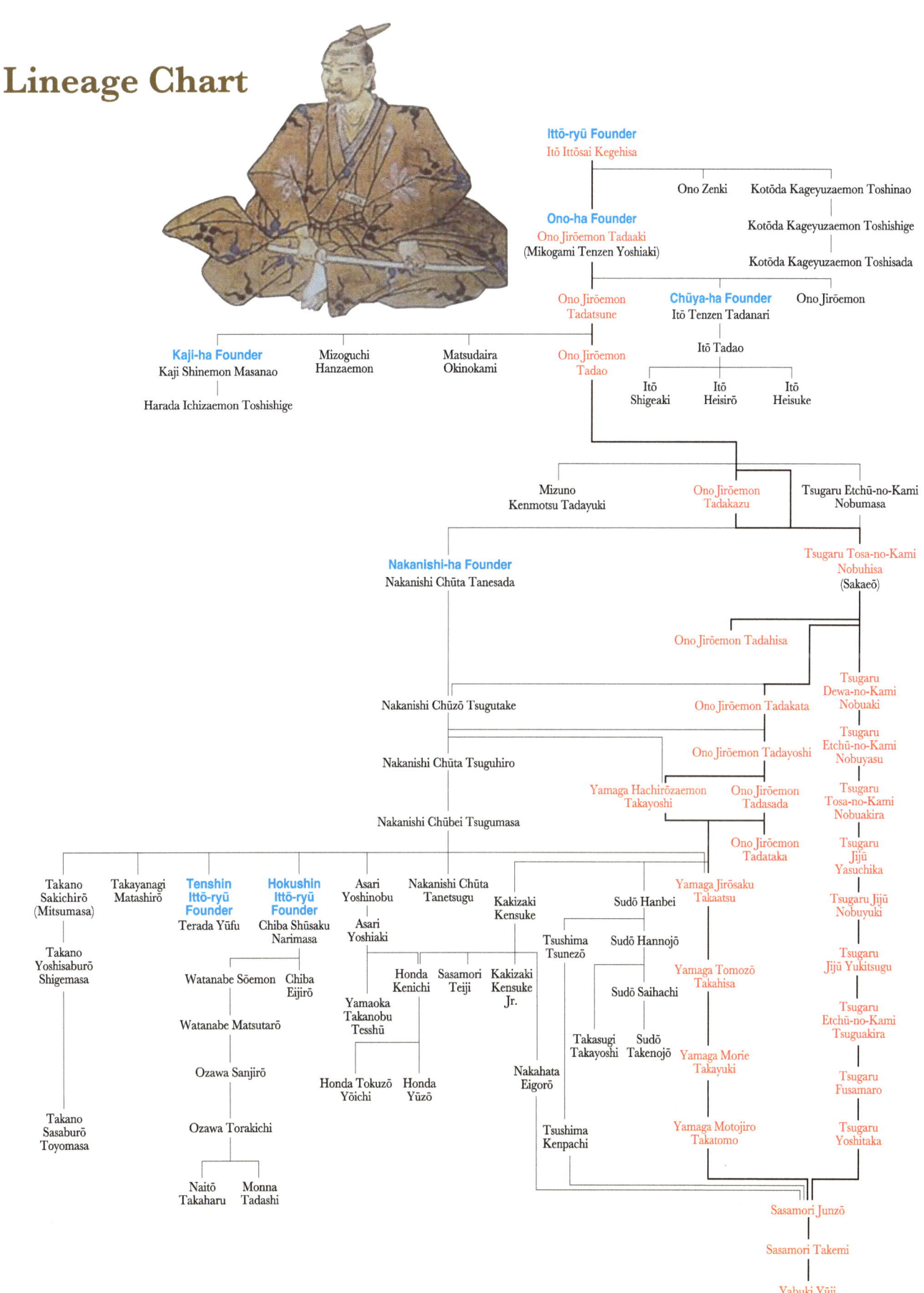

Ittō-ryū Founder
Itō Ittōsai Kegehisa

Ono Zenki — Kotōda Kageyuzaemon Toshinao

Ono-ha Founder
Ono Jirōemon Tadaaki
(Mikogami Tenzen Yoshiaki) — Kotōda Kageyuzaemon Toshishige

Kotōda Kageyuzaemon Toshisada

Ono Jirōemon Tadatsune — **Chūya-ha Founder** Itō Tenzen Tadanari — Ono Jirōemon

Itō Tadao

Kaji-ha Founder
Kaji Shinemon Masanao   Mizoguchi Hanzaemon   Matsudaira Okinokami   Ono Jirōemon Tadao   Itō Shigeaki   Itō Heisirō   Itō Heisuke

Harada Ichizaemon Toshishige

Mizuno Kenmotsu Tadayuki   Ono Jirōemon Tadakazu   Tsugaru Etchū-no-Kami Nobumasa

Nakanishi-ha Founder
Nakanishi Chūta Tanesada   Tsugaru Tosa-no-Kami Nobuhisa (Sakaeō)

Ono Jirōemon Tadahisa   Tsugaru Dewa-no-Kami Nobuaki

Nakanishi Chūzō Tsugutake   Ono Jirōemon Tadakata   Tsugaru Etchū-no-Kami Nobuyasu

Nakanishi Chūta Tsuguhiro   Ono Jirōemon Tadayoshi   Tsugaru Tosa-no-Kami Nobuakira

Yamaga Hachirōzaemon Takayoshi   Ono Jirōemon Tadasada   Tsugaru Jijū Yasuchika

Nakanishi Chūbei Tsugumasa   Ono Jirōemon Tadataka   Tsugaru Jijū Nobuyuki

Takano Sakichirō (Mitsumasa)   Takayanagi Matashirō   **Tenshin Ittō-ryū Founder** Terada Yūfu   **Hokushin Ittō-ryū Founder** Chiba Shūsaku Narimasa   Asari Yoshinobu   Nakanishi Chūta Tanetsugu   Kakizaki Kensuke   Sudō Hanbei   Yamaga Jirōsaku Takaatsu   Tsugaru Jijū Yukitsugu

Takano Yoshisaburō Shigemasa   Watanabe Sōemon   Chiba Eijirō   Asari Yoshiaki   Tsushima Tsunezō   Sudō Hannojō   Yamaga Tomozō Takahisa   Tsugaru Jijū Yukitsugu

Yamaoka Takanobu Tesshū   Honda Kenichi   Sasamori Teiji   Kakizaki Kensuke Jr.   Sudō Saihachi   Tsugaru Etchū-no-Kami Tsuguakira

Watanabe Matsutarō   Takasugi Takayoshi   Sudō Takenojō   Yamaga Morie Takayuki   Tsugaru Fusamaro

Ozawa Sanjirō   Honda Tokuzō Yōichi   Honda Yūzō   Nakahata Eigorō

Takano Sasaburō Toyomasa   Ozawa Torakichi   Tsushima Kenpachi   Yamaga Motojiro Takatomo   Tsugaru Yoshitaka

Naitō Takaharu   Monna Tadashi

Sasamori Junzō

Sasamori Takemi

Yabuki Yūji

Transmission Documents & Scrolls

In Ittō-ryū, there are four scrolls that explain the school's doctrine and expound on its secrets. These are: the Ittō-ryū Heihō Jūnikajō scroll; the Kana scroll, the Hon scroll, and the Wari scroll. Ono Jirōemon Tadaaki took the lessons he learned from Itō Ittōsai Kagehisa and recorded them in these secret scrolls.

When students aspiring to learn Ittō-ryū join the school and begin their training, they will receive the Jūnikajō scroll as the first level of transmission after they spend many years mastering its curriculum, understanding its underlying principles, and having their progress recognized by their teachers. As they devote themselves to improving even more, their minds and bodies will develop as their skills reach ever higher levels until they are awarded the Kana scroll as the intermediate level of transmission. When they diligently dedicate themselves to the Way, constantly improving their abilities, refining their mental discipline, cultivating martial virtue, and mastering all the secrets of the school, they will be awarded the Hon scroll as the third and highest level of initiation. We say that once a student is awarded the three scrolls listed above, they have received full transmission in the art, or menkyo kaiden. Those who achieve this will be further awarded the Deshi Toritate License, and once they are awarded the Keikojō License, they will be permitted to establish a new dojo or take over an existing dojo from a teacher who retires. Aside from these three documents, there is also the Wari scroll, the fourth that is considered so secret it was only passed on from the sōke to a single heir.

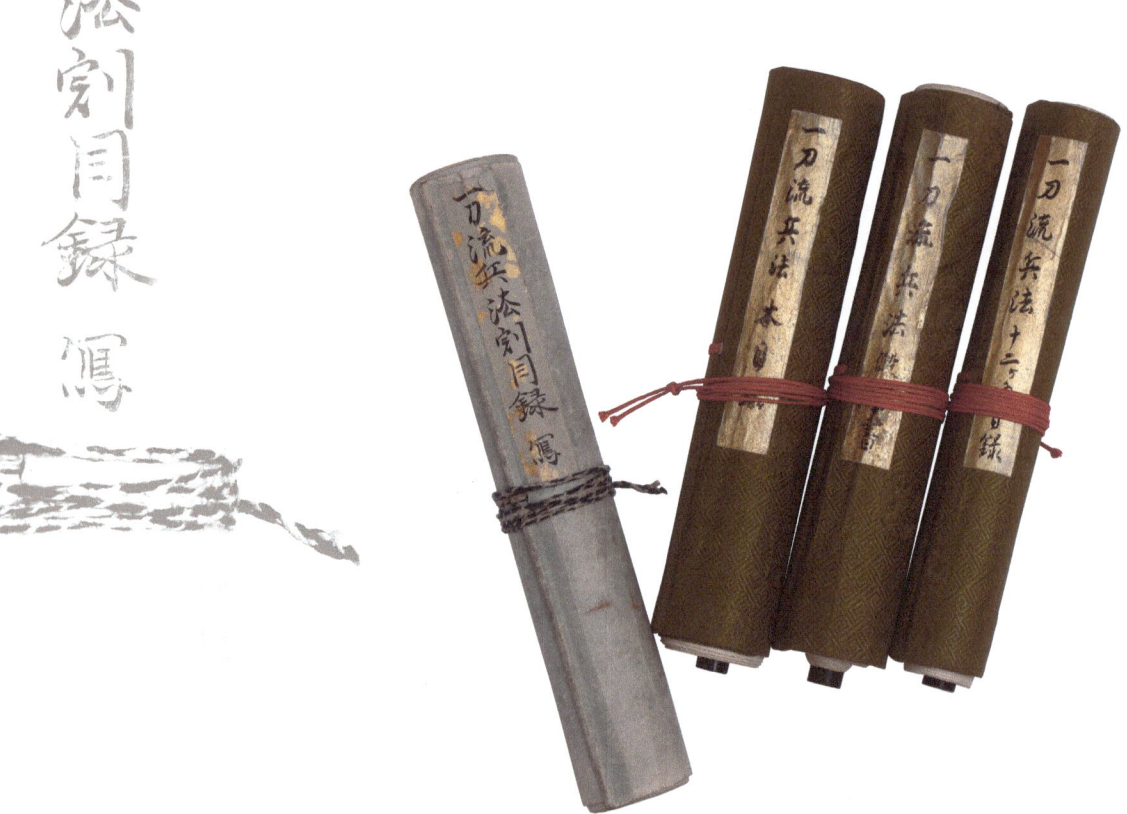

• Ittō-ryū's Transmission Documents

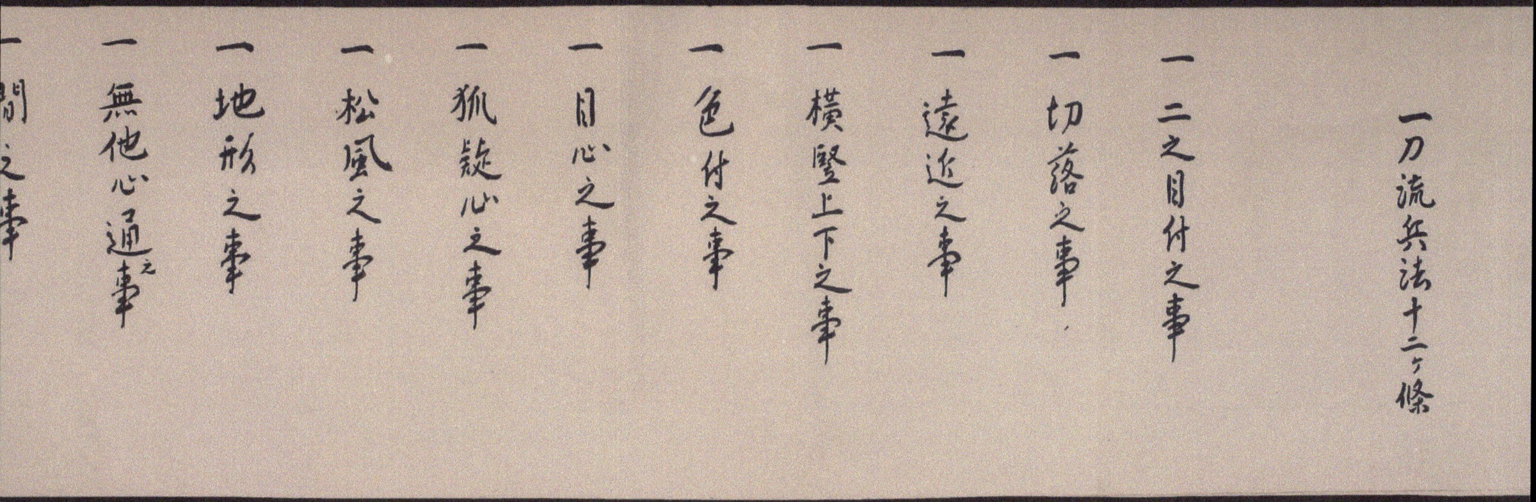

一刀流兵法十二ヶ條

一 二之目付之事
一 切落之事
一 遠近之事
一 横堅上下之事
一 色付之事
一 目心之事
一 狐疑心之事
一 松風之事
一 地形之事
一 無他心通之事
一 間之事

• Jūnikajō Scroll (initial transmission)

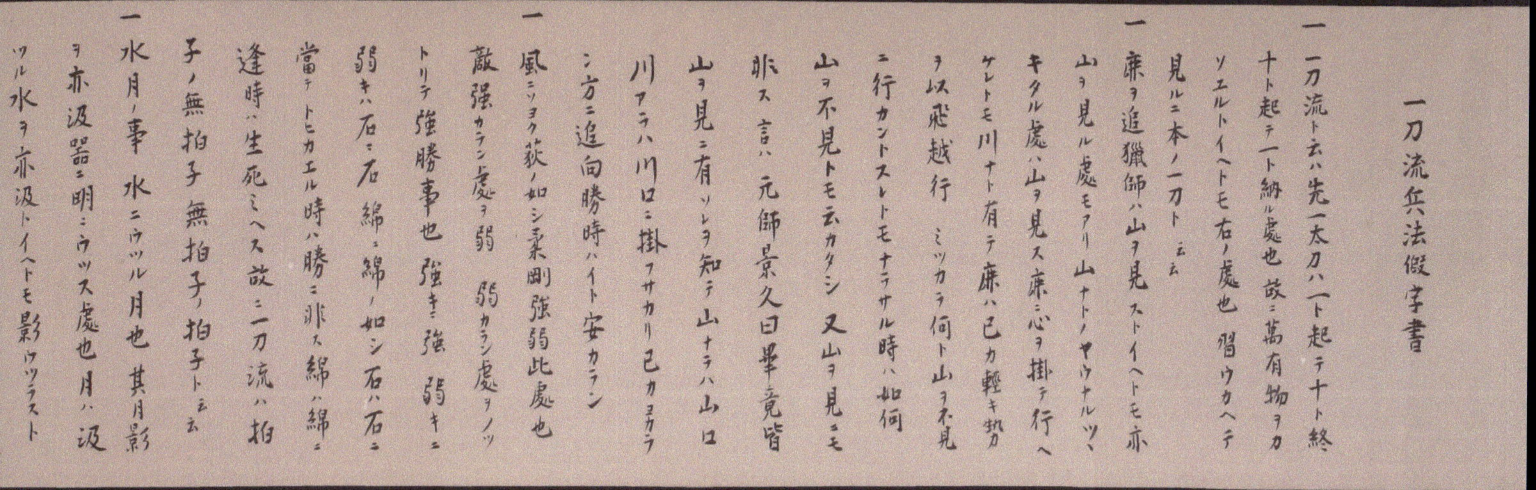

一刀流兵法假字書

一 一刀流ト云ハ先一太刀ハ一ト起テ十ト終
十ト起テ一ト納ル處也 故ニ萬有物ヲカ
ソエルトイヘトモ右ノ處也 習ウカヘテ
見ルニ本ノ一刀ニ云

一 廉ヲ追獵師ハ山ヲ見ルモアリ一山ヲ見ルトモ亦
山ヲ見ル處ナルニ山ヲ見ス廉ニ心ヲ掛テ勢ヘ
ケトトモ川ト出有テ廉ハ已カ輕キ勢カ
ヲ以飛越行 ミツカラ何ト山ヲ不見
二行カントスレトモトモナラサル時ニ如何
山ヲ不見トモ云カタシ 又山ヲ見ニミ

非ス言ハ元師景久曰畢竟皆
山ヲ見ニ有ソレヲ知テ山ナラハ山ニ山ロ
川アラハ川ロニ掛ナサカリ已ニヨカラカ
ニ方ニ追向勝時ハイト安カラン

一 風ニツヨク萩ハ如シ乗則強弱此處也
嚴強カラン處ヲ弱 弱カラン處ヲノツ
トリテ強勝事也 強キ強ニ弱キニ
弱キハ石ニ石 綿ニ綿ノ如ク石ハ石ニ
當テ石ニカエル時ハ勝ニ非ス綿ニ綿ニ
逢時ハ生死ニヘ大故ニ二刀流ハ拍
子ノ無拍子無拍子トモ云

一 水月ノ事 水ニウツル月也 其月影
ヲ汲器ニ明ニウツル處也月ハ汲
ツル水ヲ亦汲トイヘトモ影ウツラスト

• Kana Scroll (intermediate transmission)

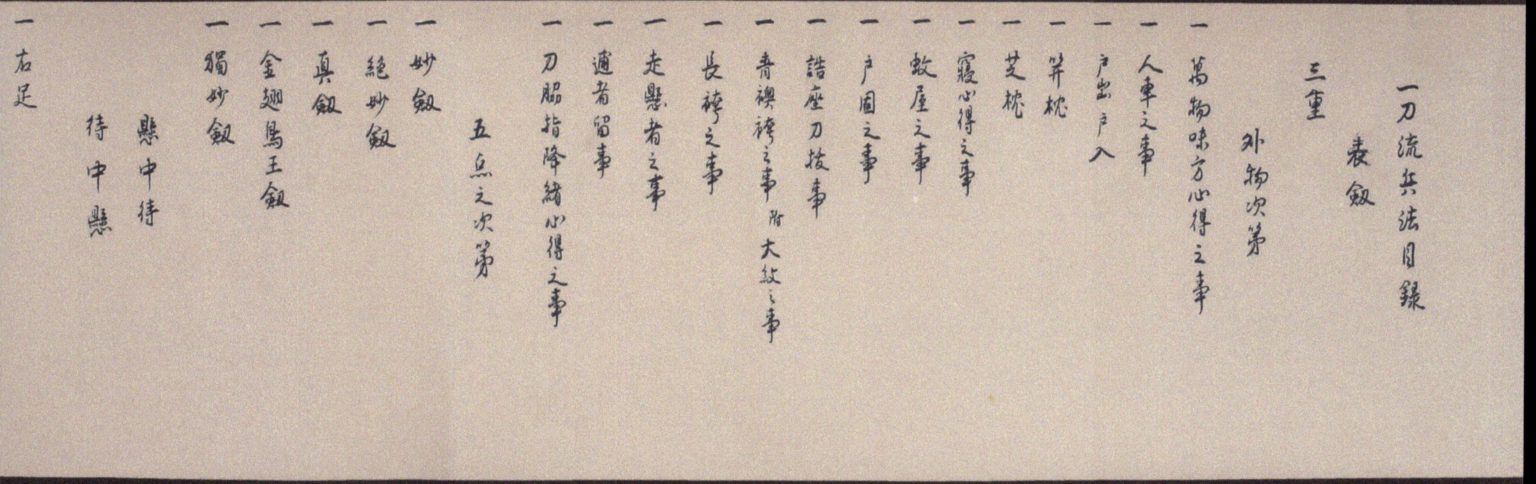

一刀流兵法目録

表劔
三重
外物次第

一 萬物味方心得之事
一 人車之事
一 戸出戸入
一 笄枕
一 芝松
一 寝心得之事
一 蚊屋之事
一 戸囲之事
一 詰座刀披事
一 寿禊禊之事 附大紋之事
一 長袴之事
一 走懸有之事
一 適着留事
一 刀脇指降緒心得之事

五点之次第

一 妙劔
一 絶妙劔
一 真劔
一 金廻鳥王劔
一 獨妙劔

懸中待
待中懸

一 右足

• Hon Scroll (advanced transmission)

Ittō-ryū Wari Scroll

The Ittō-ryū Heihō Wari scroll is a document of secrets handed down from the sōke of the mainline school to only one student, and not disseminated to anyone else, no matter how skilled they might be. Since the founding of Ittō-ryū, all sōke held this document close to their chests, but the 16th Sōke, Sasamori Junzō, a leading figure in the world of kendo, dared to expose it to the world to promote a greater understanding of the principles that underpinned the foundations of kendo, as opposed to the kendo of the time which was derisively referred to as "hit and run" kendo—an unfortunate legacy of the post-war sport of Shinai Kyogi.

The Ittō-ryū Heihō Wari scroll integrates into a single document the lessons of: Goten, Juniten, Kuka, Goka, Metsuke, and Tsuke, which are regarded as the highest-level secrets of our school. It is the most profound of all of the scrolls, laying out these lessons in an orderly, step-by-step fashion, breaking them down into their component parts and finely analyzing them so that one can easily instantiate the spiritual and philosophical, i.e. metaphysical, ideals of swordsmanship through physical techniques. Below we provide pictures of an actual Wari scroll. There are explanations of these lessons in Sasamori Junzō's book, *Secrets of Ittō-ryū*, along with a section on how to improve one's kendo.

Metaphysical: Things that cannot be understood through the experience of the senses and which are constrained by the forms of space and time. Supernatural; ideal.

Physical: Things that can be understood through the experience of the senses. The existence of shapes manifested in the world assuming the basic forms of space and time.

[Translation]

Day this license is issued

The Day of the Tiger in the first month; the Day of the Rabbit in the second month; the Day of the Dragon in the third month; the Day of the Snake in the fourth month; the Day of the Horse in the fifth month; the Day of the Sheep in the sixth month; the Day of the Monkey in the seventh month; the Day of the Rooster in the eight month; the Day of the Dog in the ninth month; the Day of the Boar in the tenth month; the Day of the Rat in the eleventh month; and the Day of the Ox in the twelfth month.

Goten of Shin

Myōken
   Win When Your Opponent is About to Cut Your Shoulder
   Cutting The Same Time Your Opponent Cuts
   Just Cut In
   Duck Under When Pressed
   Strike as Soon as Your Opponent Rushes In
   Tactic of Attacking Against a Jōdan
   Strike Gedan from Jōdan
   Tactic of Striking as You Move to Chūdan
   Tactic of Winning While Standing Up

Zetsu Myōken
   Noshi Uchi
   Attack Toward the Left Jōdan
   Attacking the Ten-no-Yoko Posture
   Attacking the Five Positions of Heaven
   Tactic of Slipping By to the Left or Right
   As Soon as Your Opponent Attacks, Pluck His Sword Away as in Kodachi
   When Your Opponent is in Yoko Seigan, Cut Up into his Left Hand
   When You Attack Your Opponent from the Right and He Pulls Back to Strike and You See an Opportunity to Win, Adjust Your Sword

Shinken
   Winning by Doing Kiriotoshi to the Right Side of the Opponent
   When Your Opponent Assumes Jōdan, Use the Same Tactic
   Beating Your Opponent by Standing Your Sword Up After Kiriotoshi
   Beating Your Opponent by Moving to the Left After Doing Kiriotoshi
   When Your Opponent Withdraws, Immediately Apply Yoko Seigan
   When Your Opponent Pulls Back to Yo, Apply Yoko Seigan
   React to Your Opponent's Wakigamae by Applying Seigan

Kinshichō Ōken
   Kiriotoshi from a Low Seigan
   Jōdan of Gedan
   Tactic of Cutting Through the Tangle; Take One Step Forward and Cut

Doku Myōken
   Where One Exists
   One Step, One Cut
   Apply Hongaku to Jōdan
   Apply Hongaku to Onken
   Apply Hongaku Against Hassō
   When You Use Tōhō Against Hongaku, Immediately Cut the Hands

Goten of Sō

Myōken
   When Your Opponent Approaches in Wakigamae, Approach in Seigan and Close from Below

Zetsu Myōken
   When the Enemy Approaches in Seigan, Sidestep to the Right in Wakigamae and Attack Upward from Gedan

Shinken
   When the Enemy Uses Ten no Yokogamae, Respond with Seigan

Kinshichō Ōken
   When Your Opponent Uses Hassō, Respond by Using Seigan

Doku Myōken
   When Your Opponent is in Jōdan, Take Up Seigan

Twelve Points

1. Makikaeshi
2. Aiseigan, Immediately Slip Toward the Right, Shift to Gedan, Cut Toward the Shoulder
3. From Aiseigan, Deliver a Crushing Blow from Above
4. Do Nobeshikiuchi from Aiseigan
5. Apply a Hongaku Against a Left Onken
6. When the Opponent Cuts Randomly, Circle to the Left and Approach
7. Do Saumakuri Against Uchitachi's Left Kamae
8. Against Hassō, Move from a Left-Foot-Forward Kasumi-karasu
9. Do a Horizontal Cut from Onken Against a Hassō
10. From Onken, Cut Upward from Below Against an Opponent's Jōdan
11. Move Back from Aiseigan to Onken and Cut Down Your Opponent
12. Stopping with Hotsu

The New Goten of Shin

Myōken
   There Is a Tactic of Approaching Uchitachi's Jōdan in Wakigamae and Slipping Out to His Rear

Zetsu Myōken
   Whether the Opponent Is to the Left or Right, Slip by Him in Seigan

師弟契約之日取

正月寅 二月卯 三月辰 四月巳

五月午 六月未 七月申 八月酉

九月戌 十月亥 十一月子 十二月丑

伊藤一刀齋流別目録之次第

真之五點

一 妙劍

一 からふから落るゝねの勝

一 おきよ打るゝかを

一 ちきよ打るゝを

一 押かけられてるゝをあを

一 ちきよ押ゑらちきよ打るゝをあを

一 上腹ふゑにくるゝを望

一 下腹と上腹にてるゝを川るゝを

一 中腹ふ引るゝを川るゝを

Shinken

Invite Your Opponent to Move Toward the Mist

Kinshichō Ōken

Face in Seigan an Opponent Who Is in Jōdan,
Drop the Sword and Pull Back to Jōdan for
Another Strike

Doku Myōken

When Uchikata Dashes Forward to Cut Your
Forearm, Cut His Upraised Wrist; Can be Done
from Lower, Middle, or Upper Positions

Five Sun Willow Branch

If you hold up wood from a willow tree that is powerful
and sacred, such as that used for a torii gate, it will
summon.

[Spell]: Un Da Gi Un Sha Kun Shi Tsu Chi

Oral transmission: If you use this method to summon a
woman, it will lose its power afterward.

Nine Swords

Tsumeiri
Soegiri
Mi no Kane
Randome
Yorikiri
Shin no Shinken
Saten
Uten
Shin no Seigan

Metsuke

Discarding Metsuke
Four Functions of the Sword (supplemented by oral
teachings)
Eightfold Metsuke

Attachment

Taisen
Dōchū
Shōhon
Aspects of the Four-Direction Sword

Secrets of the Five Items

Ken-no-Dan
Sword of No-aspect
Estimating Measurements
First Level
The True Kinshichō Ōken

The sword skills described above are the arts of war.
They are thought to have started with the pursuit of
bladed weapons, like the hoko, and are associated with
an attachment to Marishiten.

There were many such traditions throughout
Japan, and Itō Ittōsai Kagehisa, of the Ōmi Domain,
studied several of these schools of strategy far and wide.
Later, after contemplating the true nature of victory, he
realized that advantage lay in neither the long nor the
short, but in the ground in between, and this is what he
came to teach. He taught his students so they could
immediately grasp that it didn't matter if their weapons
were short or long, they could still succeed with either.
All that matters is overwhelming one's opponent, like
towering reeds growing near a spring. With that in
mind, he selected techniques from traditions of bygone
times and established his own school. Hoshatō is living
and dying, conferring life and dealing death using the
Four Cuts. Those who master the Ultimate Secrets
surpass all others in this world, are rare, and hardly
ever appear. It's obvious that people don't know what
others say and don't say. The attitude of everyone in
this world toward the ancient teachings is shallow.
Their understanding of these is akin to trying to break
a rock with an egg, twisting metal, or heaping praise on
a pile of rubble. Who would purposely spend their
precious moments doing such useless things with such
little time to waste? Shouldn't they serve a higher
purpose?

Ultimate Secrets

Hoshatō
Two Directions, One Step
Eight Directions, Three Steps
Left Foot
Right Foot

Four Cuts

The transmission of Ittō-ryū Heihō to you is highly
fitting and reasonable. Learning how the weak controls
the strong and the soft controls the hard is the means to
acquire this understanding and become victorious.
There have been many, many occasions when you
engaged in discussions and explanations and your
efforts are on par with that of a master. For these
reasons, I present you with this Wari scroll. This
completes the transmission to you of all four scrolls
from our school from beginning to end, leaving nothing
out. From this day forward, please strive to fulfill the
lessons contained herein by devoting yourself to further
honing your skills even more. Thus, there is nothing
more to add.

Itō Ittōsai Kagehisa (founder)
Ono Jirōemon Tadaaki
(Others omitted for brevity)

Ittō-ryū Heihō Kumisū Scroll

This document was sent to Yamaga Hachirōzaemon Takayoshi when the 7th Sōke, Ono Jirōemon Tadasada, received a license.

After that, it was passed down from the House of Yamaga to Sasamori Junzō, then to his third son, the 17th Sōke, Sasamori Takemi, and then to Yabuki Yūji, the 18th and currently serving Sōke. Sasamori Junzō was born as the sixth son of Sasamori Yozo, a retainer of the former Hirosaki Domain. Junzō entered the Hokushindo Dojo at the age of eight, learned the Ono-ha Ittō-ryū that was taught within the Hirosaki Domain at the time, and later studied kendo at Waseda University under Takano Sasaburō. Later on, he would become the 16th Sōke of Ono-ha Ittō-ryū and inherit both Shin Musō Hayashizaki-ryū Iajutsu and Chokugen-ryū Naginata-jutsu. The original documents verifying this are held by the Reigakudō.

Because this scroll is quite long, we are presenting the last part of it and the section showing the first five moves of the kumitachi along with their contemporary interpretations.

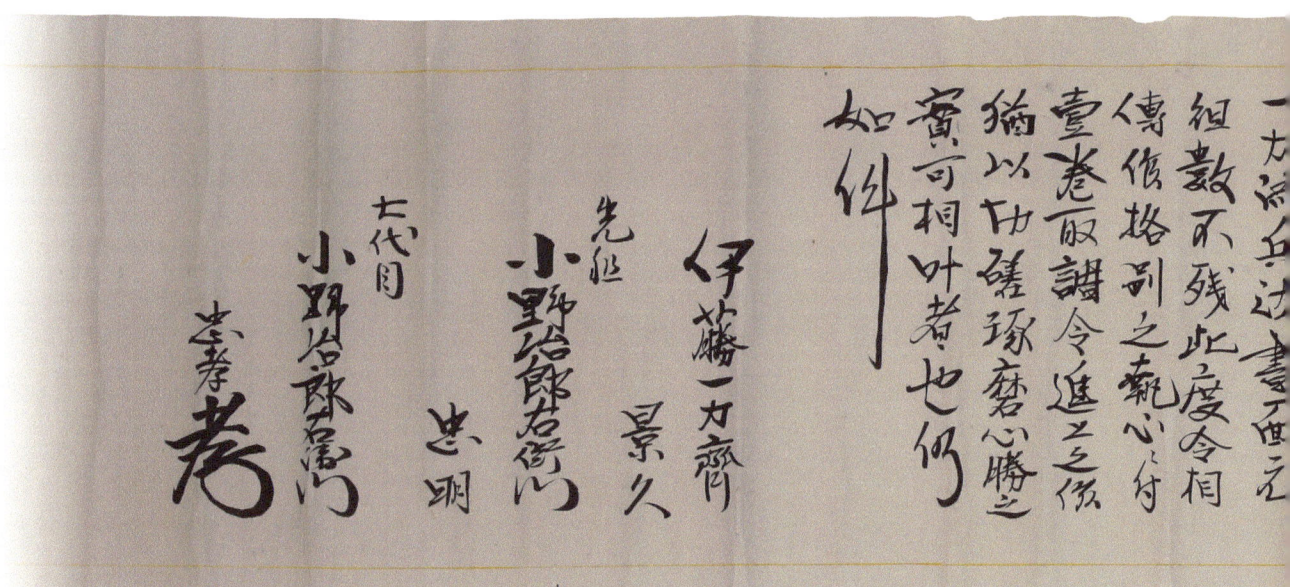

Hitotsugachi
Uchitachi starts in intō, while Shikata starts in seigan and does kiriotoshi. Shikata cuts Uchitachi's left forearm for the victory and then shifts to gedan.

Nihonme
Uchitachi starts in seigan while Shikata starts in gedan and then shifts to seigan. Shikata thrusts towards Uchitachi's men and then brings his sword around to cut his right forearm for the win. Shikata then shifts to jōdan.

Sanbonme
Uchitachi approaches in intō opposite Shikata, who is in seigan. Uchitachi cuts towards Shikata's hands. Shikata drops his sword and then brings it around to strike Uchitachi's right forearm. Shikata then shifts to jōdan.

Gedan-no-Kasumi
While Uchitachi is in a gedan-kasumi kamae, Shikata shifts from gedan to touch sword tips. When Uchitachi brings his sword around to cut, Shikata cuts his hands. Shikata then cuts Uchitachi's right forearm and shifts to gedan.

Wakigamae-no-Tsuke
Uchitachi approaches in seigan opposite Shikata, who is in a wakigamae stance. Shikata shifts to seigan and touches sword tips with Uchitachi. Both sides push in toward each other. Uchikata throws off Shikata's sword and Shikata brings his sword around and cuts Uchitachi's right forearm for the win. Shikata then shifts to jōdan.

一刀流兵法組数目録

一二の腮
打太刀陰ニ抖蝶之方清眼ニ当而
し右へしつゝ切腮下ニ而取位

一二本目
互方清眼筈ニ方古ゟり清眼付
打太刀面ニ実込込ゟ右へしつゝ切
腮上ゟ二車信

右之組　五十
都合大刀　数五拾本

外ニ

一功落シ
一ツゝミ
一ハリ
一懸意

一小太刀
一柏小太刀
一三重
一ハリ一

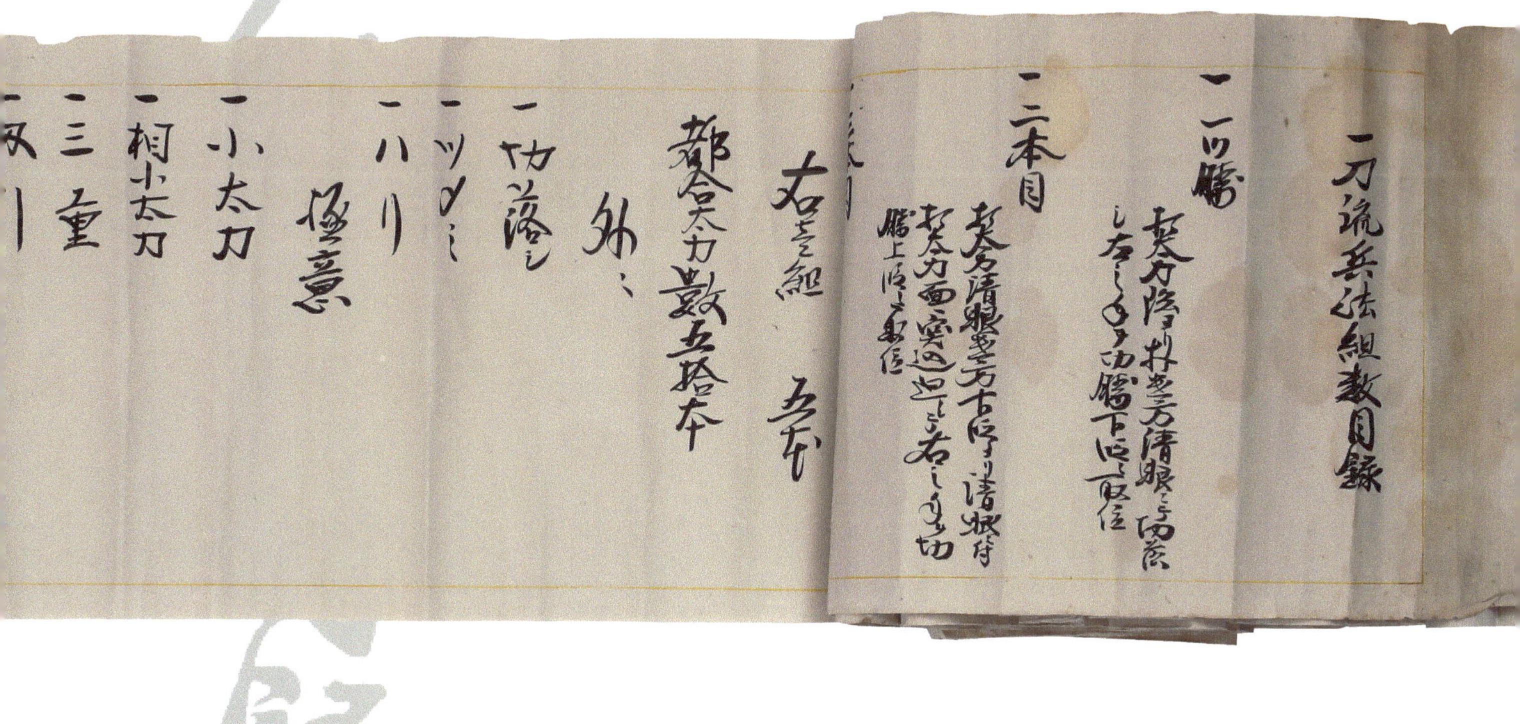

Connection to Modern Kendo:
a "Unified Way of the Sword" that Transcended Schools and Factions

In April, 1895, the Dai Nippon Butokukai was established in Kyoto. It began as a conglomeration of individual kenjutsu schools and factions. Each of the factions had their own kata and individual methods of training, which they variously called *kenjutsu* and *gekken*.

To streamline kendo education for public school students, however, a research committee to develop a new set of kendo kata was established in the Butokukai in 1912. Twenty-five committee members were selected nation-wide, with five chief examiners forming the nucleus of those who would develop the detailed moves of the kata. These chief examiners were: Takano Sasaburō (Ono-ha Ittō-ryū [Nakanishi-ha]), Naitō Takaharu (Hokushin Ittō-ryū), Monna Tadashi (Hokushin Ittō-ryū), Negishi Shingorō (Shintō Munen-ryū), and Tsuji Shinpei (Shingyōto-ryū, Jikishinkage-ryū). At the time, it was thought that creating a common kata that transcended the various schools and their factions would not be easy.

They say that Takano Sasaburō concealed a small knife in his sleeve and approached his duties with such determination that he was prepared to die if he didn't get his way, but in actuality, it seems that only the most basic techniques were selected for teaching middle and high school students. In October, 1912, the Dai Nippon Imperial Kendo Kata (current Japan Kendo Kata) were standardized. Of the five chief examiners who approved the kata, three were from an Ittō-ryū line, and despite the fact that it does not contain kiriotoshi, the movements within the kata appear to have been strongly influenced by the techniques of Ittō-ryū.

Takano Sasaburō
July 9, 1862 – Dec 30, 1950

Japanese kendoka. Kendo Hanshi of the Dai Nippon Butokukai. Worked as a metropolitan Police Department Instructor of Gekken, professor at the Tokyo Higher Normal School, among others. A leading figure in the world of kendo at the beginning of the Shōwa Period. Sasamori Junzō (later to become a Minister of State and the 16th Sōke of Ono-ha Ittō-ryū) was a student of his in the Waseda University Kendo Club. Sasaburō's grandfather, Sakichirō (Mitsumasa) was a student of Nakanishi Tsugumasa, the 4th head of the Nakanishi-ha Ittō-ryū, and worked as the official kenjutsu instructor to the Okudaira Shohei camp, the main family of the Oshi Domain. Nakanish-ha Ittō-ryū was itself called Ono-ha Ittō-ryū, and Takano himself also referred to it as Ono-ha.

Naitō Takaharu
December 16, 1862 – April 9, 1929

Japanese kendoka. Sixth son of Ichige Gorōemon Takanori, archery instructor of the Mito Domain, he took his adopted family name of Naitō. Hailed from the Hokushin Ittō-ryū. Held the title of Kendo Hanshi in the Dai Nippon Butokukai. Professor at the Budo Senmon Gakkō. Along with Takano Sasaburō, had a huge influence on the world of kendo, where they used to say, "Naitō of the West; Takano of the East." At the age of twelve he entered the Tōbukan Dojo of Hokushin Ittō-ryū master Ozawa Torakichi and later worked as a Senior Professor of Kendo at the Bujutsu Kyōin Yōseijo (later renamed the Budo Senmon Gakkō). In 1911, he was appointed as one of the chief examiners to the committee that standardized the Dai Nippon Imperial Kendo Kata. Worked as a Hanshi at the Tokyo Senmon Gakkō (currently Waseda University) Gekken Club, and also associated with Sasamori Junzō, the 16th Sōke of Ono-ha Ittō-ryū.

Ogawa Kinnosuke

May 18, 1884 – March 30, 1962

Japanese kendoka who achieved the rank of Hanshi 10th Dan. Born in Iwakura Village (now Iwakura City), Niwa County, Aichi Prefecture. Departed for Nagoya and entered the Hokushin Ittō-ryū school of Katō Kanichi (foster father of Katō Shichizaemon). Appointed as an assistant instructor at the Budo Senmon Gakkō in 1914 at the request of Naitō Takaharu. After the passing of Naitō, was appointed as the principal instructor of the Budo Senmon Gakkō, where he worked until 1944.

Saimura Gorō

May 4, 1887 – March 13, 1969

Japanese kendoka who achieved the rank of Hanshi 10th Dan. Famed as one of the Sword Saints of Shōwa, was known as "Sword Saint 10th Dan." Born the third son of Saimura Kasei, a retainer of the former Fukuoka Domain in Yōhano-chō, Fukuoka City (present day Daimyō-chō). Started studying kendo in earnest at the defunct Aidō Dojo of Yoshitome Katsura (Abe-ryū kendo master and Black Ocean Society member) and at the prestigious Fukuoka Prefectural Shuyukan Middle School. Was among the first cohort to enter the Dai Nippon Butokukai Bujutsu Kyōin Yōseijo where he studied under Naitō Takaharu. After the war, he worked tirelessly to revive the kendo that had been banned by U.S. military occupation authorities. Was awarded his kendo 10th Dan in 1957 from the All-Japan Kendo Federation. He was awarded the title of honorary Kendo Shihan from the Tokyo Metropolitan Police Department and Professor Emeritus from Kokushikan University.

Sasamori Junzō

May 18, 1886 – February 13, 1976

Japanese politician, educator, kendoka. Imperial honors: the Order of the Sacred Treasure, First Class, Third Degree. Titles: Doctor of Philosophy (Ph.D.), Kendo Hanshi. Professional positions held include: Chancellor, Too-Gijuku; Chancellor, Aoyama Gakuin; Member of Parliament, House of Representatives (4 terms); Member of Parliament, House of Councillors (3 terms); President of the Demobilization Agency; Director General of the Repatriation Agency; and Chief Advisor to the All-Japan Kendo Federation. After studying the Ono-ha Ittō-ryū that had been passed down within the Hirosaki Domain at the Hokushindo Dojo, he studied kendo in the Waseda University Kendo Club under Takano Sasaburō. Later became the 16th Sōke of Ono-ha Ittō-ryū and also inherited the schools of Shin Musō Hayashizaki-ryū Iaijutsu and Chokugen-ryū Naginata-jutsu.

Correspondence from Takano Sasaburō to Sasamori Junzō

The collection contains correspondence dated June 8, 1925 that is addressed to Sasamori Junzō from Takano Sasaburō, who at the time was at the Kudan Meishinkan Headquarters of the Shūdōgakuin. "In order to honor the 300th anniversary of the death of the founder of Ono-ha Ittō-ryū, a request has been made to invite Nakahata Eigoro Sensei." 300 years before 1925 was 1625 when Tadaaki founded Ono-ha Ittō-ryū at the age of 57 (he died three years later). So, does this suggest that Ono-ha Ittō-ryū was created when the 2nd Sōke, Tadatsune, was around 17 years old? Also, Nakahata Eigoro took over the former Hirosaki Domain's instruction of Ono-ha Ittō-ryū and was Sasamori Junzō's teacher. Moreover, in 1925, Sasamori Junzō, together with Ichikawa Umon and Shibuya Fumio, were setting up the Gokokukan Dojo in Hirosaki City.

The sentence in the text that reads, "These days, there is no one other than Sensei who is doing the kata of our old school well" abounds with a feeling of respect toward Nakahata Eigoro. The mainline of Ono-ha Ittō-ryū was maintained by both the Tsugaru and Yamaga families, but it seems that, in reality, the instruction was conducted by the private students in each of their respective dojos. Nakahata received initiation by both the Nakanishi and Yamaga, and was a direct student of Kakizaki Kensuke, who served the former Hirosaki Domain, and seems to have taught his students. Takano Sasaburō referred to his own style as Nakanishi-ha Ittō-ryū from the line his grandfather, Sakachirō Mitsumasa, studied, but there are also documents in which he refers to what he did as Ono-ha Ittō-ryū.

• 1903 Demonstration at the Too-Gijuku (Left: Takano Sasaburō, age 41; right: Nakahata Eigoro, age 57)

1. Move 2, Mukaizuki: Shikata is about to stab toward Uchikata's solar plexus.

2. Move 5, Wakigamae-no-tsuke: right after Shikata shifts to wakigamae from sokuizuke.

3. Moves 8 & 9, Intō: just as Shikata is about to do the zig-zag step.

[Translation]

June 9

Sasamori Sensei,

I cannot apologize enough for taking so long to write.
Although the heat still remains, I hope you are healthy and in good spirits in both mind and body.
I am fortunate to be doing fine.
Well, of course, Nakahata Sensei is as energetic as always, and seems to enjoy the kata and other activities. These days, there is no one other than Sensei who is doing the kata of our old school well. I would like to ask him to come to Tokyo once. If that is possible, I would like to hold an event commemorating the 300th anniversary of our founder, Ono [Tadaaki], in Chiba Prefecture on September 28 of this year, and I want to ask him to be my demonstration partner. I would like him to consider it for the sake of Ono-ha Ittō-ryū, and I would like you to ask him what he thinks about this. If he readily accepts, I will start my preparations.
I am very sorry to bother you while you are busy, but I appreciate your support.
Let me know if there is anything I can do for you.

Signed: Takano Sasaburō

There is another letter dated November 11, 1926 and addressed to "Sasamori Sensei," i.e. Sasamori Junzō. Due to Takano's planned business trip to open a dojo in Tohoku University, it is a long letter expressing his gratitude regarding his appointment to Waseda University Kendo Club and his wish for continued support. There was a big age difference between them—Sasamori was born in 1886 and Takano in 1862—but they were bound together through the instruction Takano provided Sasamori at the Waseda University Kendo Club. However, this letter shows the deep respect that Takano had for him.

In the middle of the Meiji Period, when Sasamori Junzō was around seven or eight years old, he received his first initiation into Ono-ha Ittō-ryū at Hirosaki City's Hokushindo Dojo at the hands of the likes of Tsushima Kenpachi, the official kenjutsu instructor of the former Hirosaki Domain. When he was a teenager, he continued to learn Ono-ha Ittō-ryū from Tsushima's successor, Nakahata Eigoro, who continued to teach unabated until the ripe old age of 82. Also, Yamaga Sokō's 4th generation descendant, Yamaga Hachirōzaemon Takayoshi, received full transmission of the school from Ono Jirōemon Tadayoshi, the Yamaga family having served the Tsugaru Domain for generations as teachers of both military strategy and Ono-ha Ittō-ryū Heihō. Takayoshi's 4th generation descendent, Yamaga Takatomo, passed on to Sasamori Junzō all of the family's tomes related to Ono-ha Ittō-ryū and Yamaga Takaatsu's treasured wooden sword that was made in 1767. Moreover, Junzō was directly bestowed in full the mainline Ittō-ryū that was taught generation after generation within the House of Tsugaru from the time it was handed down from Ono Jirōemon Tadakazu to Tsugaru Tosa-no-kami Nobuhisa, thus receiving the entirety of the techniques of Ittō-ryū, its scrolls, oral teachings, and other written commentaries on its esoteric secrets. This material also suggests that his relationship with Takano Sasaburō transcended that of mere teacher and student.

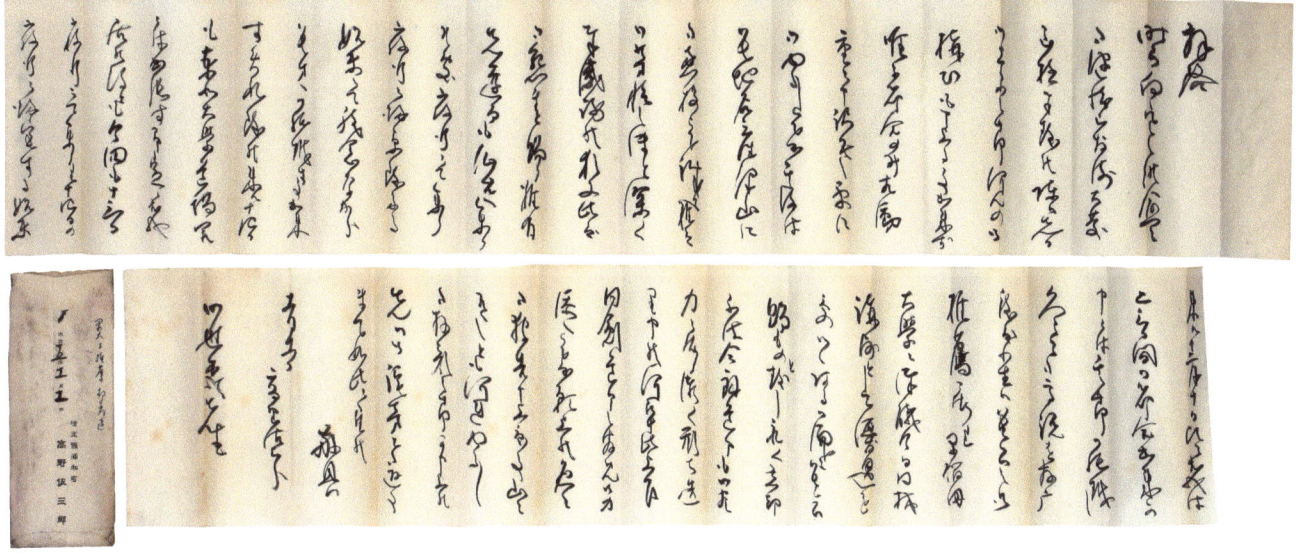

[Translation]

November 11

Sasamori Sensei,

I hope you are doing well as we approach the cold of winter.
I am deeply sorry for not being able to provide any hospitality during your recent travel to Tokyo merely due to my own convenience. Moreover, I would like to sincerely thank you for the many local specialties you provided. The other day I went to Sendai, but unfortunately, I could not visit you because I had to take an overnight train back to Tokyo. On the 14th, I will make a trip to open a dojo in Tohoku University, but I will depart Tokyo on the night of the 13th and return the night of the 14th. I will be free for two or three days around December 10th. On that occasion, I would like to hear your considered opinion after not seeing you for such a long time. After receiving your recommendation, I was appointed to the position at Waseda University. I owe the warm reception I received there all to you. This is something I will never forget. The Kendo Club is also shaping up thanks to your hard work. I look forward to your continued support in this endeavor.
There is a lot I would like to report, but I will do so in person at some point. But first, I wanted to apologize and reply.

Signed: Takano Sasaburō

Correspondence from Naitō Takaharu to Sasamori Junzō

This is correspondence from Naitō Takaharu to Sasamori Junzō, dated April 9 (year not specified). It is clear from this that Naitō was thanking Sasamori for his letter and that that there was an enthusiastic exchange of opinion regarding topics of the martial arts world on which they were in agreement. Naitō, who had a close relationship with Sasamori, was working at the time as an instructor at the Tokyo Senmon Gakkō (now Waseda University) Gekken Club, where he was teaching the students there the basics of Ittō-ryū.

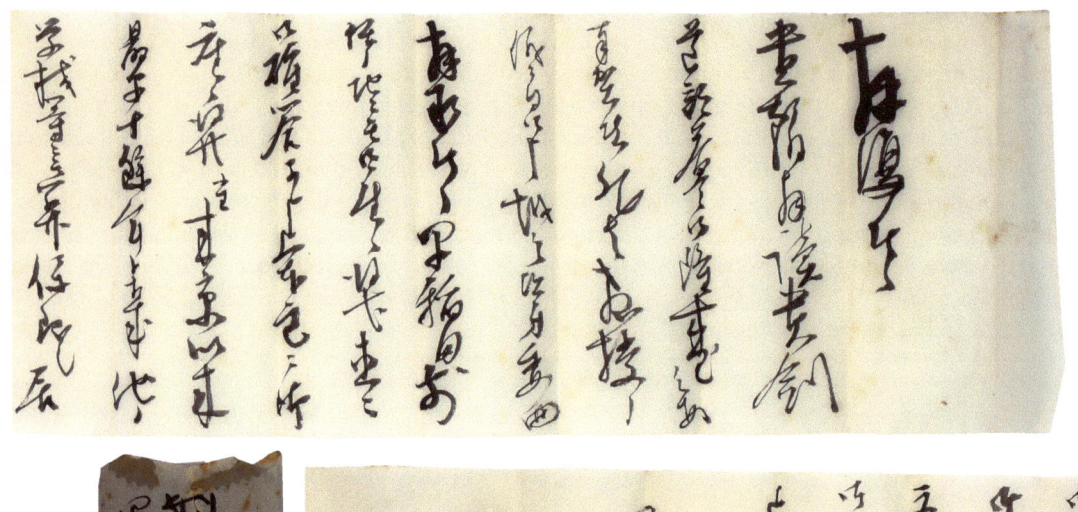

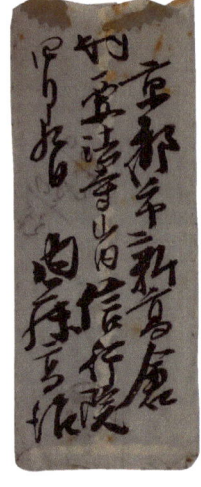

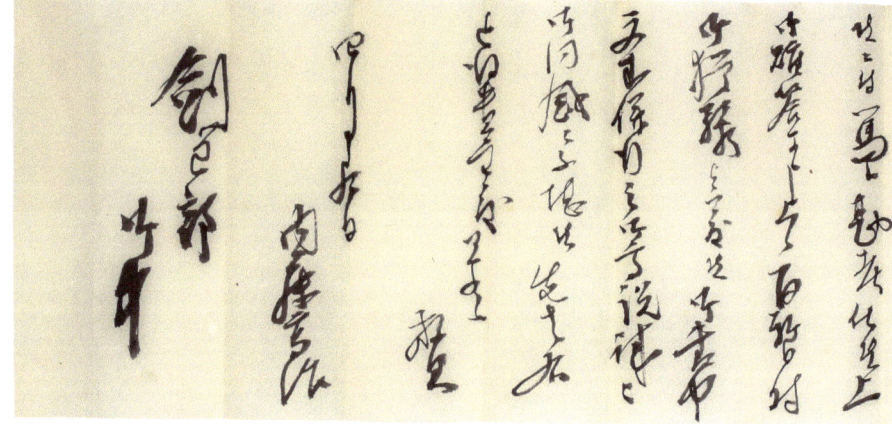

[Translation]

April 9

Kendo Club

Please allow me to provide my reply.
I am pleased to read in your letter that your kendo club is thriving. Congratulations.
I reviewed in detail the issue about the instructor position.
I will try to get back to you soonest as this matter is in regards to my previous position at Waseda.
I have been in Kyoto now for ten years and have a relationship with another school which I have to seriously consider. I want to respond with a definite answer, so please give me some time.
I completely share your sentiment about doing the military and literary arts in parallel as presented in your letter.
I will take your request under serious consideration.

Signed: Naitō Takaharu

Correspondence from Ogawa Kinnosuke to Sasamori Junzō

Ogawa Kinnosuke was born in 1884 in Iwakura Village, Niwa-gun, Aichi Prefecture (present day Iwakura City). He became a student of Katō Kanichi of the Hokushin Ittō-ryū in Nagoya. While serving in the Aichi Prefecture police force, he became an instructor at the Aichi Prefecture Police Academy from which he was recruited by Naitō Takaharu to be an assistant instructor at the Budo Senmon Gakkō in 1914. After the passing of Naitō, he became a full instructor there. In his August 18th correspondence sent to Sasamori Junzō, he expressed his gratitude for various kinds of assistance when traveling to Sasamori's local branch tournament as well as his huge help when visiting the Gokokukan dojo.

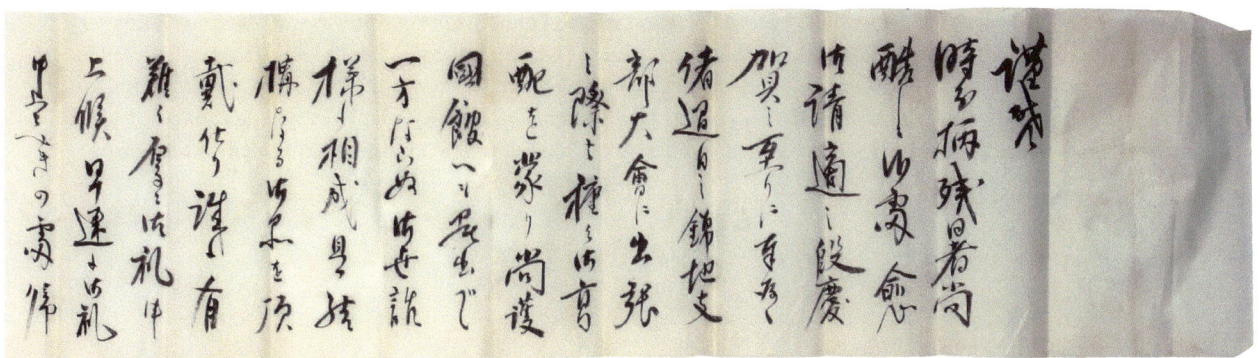

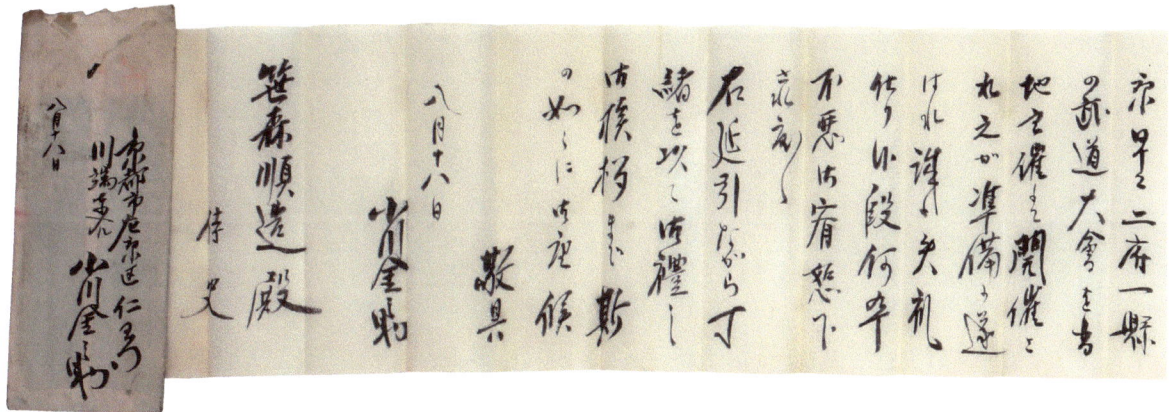

[Translation]

August 18

To: Mr. Sasamori Junzō

The lingering heat is severe, but I am happy that you are doing well.
I would like to thank you for everything you did for me during my recent business trip to your branch tournament, and for my visit to the Gokokukan. I appreciate your exceeding kindness. Moreover, thank you very much for the splendid gift. I should have expressed my appreciation sooner, but I was busy preparing to host the Kyoto-Osaka-Hyogo Martial Arts Competition as soon as I returned to Tokyo. Please forgive me.
This is my way of saying thank you.

Signed: Ogawa Kinnosuke

Correspondence from Saimura Gorō to Sasamori Junzō

Saimura Gorō was born in 1887 in Fukuoka City, Fukuoka Prefecture. He studied under Naitō Takaharu as part of the first cohort accepted in the Dai Nihon Butokukai Bujutsu Kyōin Yōseijo. After graduation, he became an assistant instructor of kendo at the Budo Senmon Gakkō (formerly called Bujutsu Kyōin Yōseijo). Below is correspondence dated October 26 addressed to Sasamori Junzō where he is referred to as "Sasamori Sensei." This correspondence describes in detail such things as his appreciation for the hospitality Sasamori extended to the Waseda University Kendo Club members he was leading, his hope for the future of kendo, etc. The Mr. Ichikawa referred to in the text is Ichikawa Umon who hailed from the same hometown as Sasamori Junzō. He also came from the same Bujutsu Kyōin Yōseijo as Saimura Gorō and was his close friend. Ichikawa later became a kendo instructor at the former Second Higher School (now Tohoku University), a kendo instructor at the Aomori Prefectural Police Academy, and a kendo instructor at Aomori Junior High School. Moreover, the Greater Youth Association referred to in the postscript was a conservative think tank centered on the Chikuzen Student Association, at the time made up of students from Fukuoka who were graduates of the Waseda University Debate Society, Kendo Club, and Judo Club. Saimura Gorō was the honorary chairman. Shibata Tokujirō (from Fukuoka), the founder of Kokushikan, was also a graduate of Waseda University. Later, Saimura Gorō was invited to become a kendo instructor at Kokushikan Senmon Gakkō, where he trained many talented kendoka.

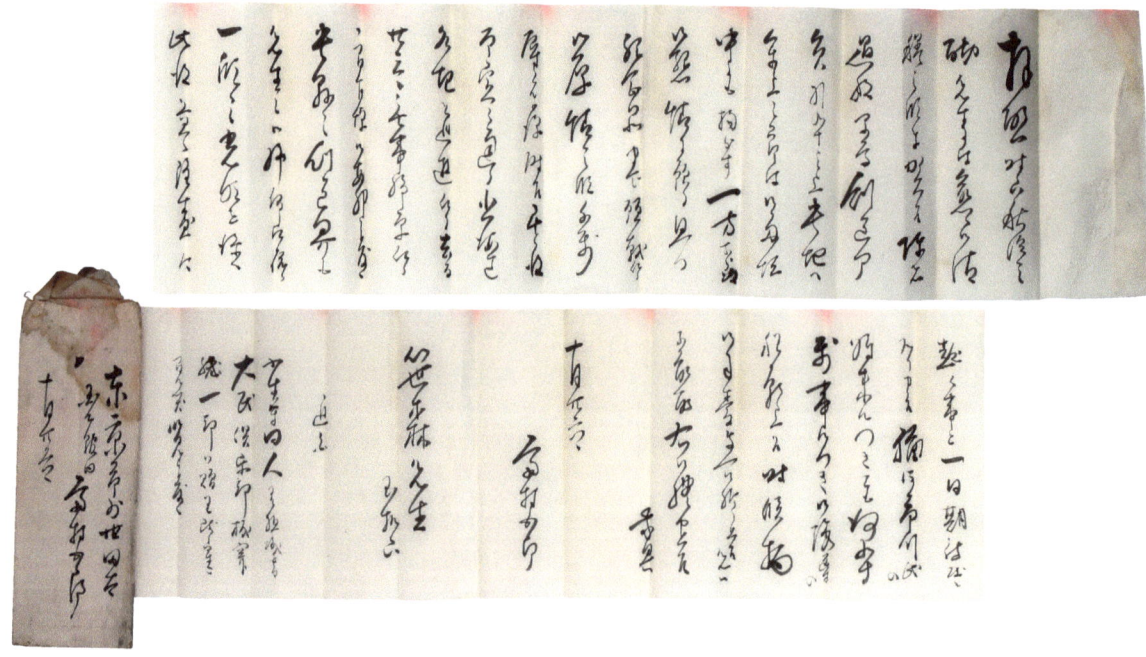

[Translation]

October 26

Sasamori Sensei,

It has gotten colder as autumn approaches, but I am happy nothing has changed with you.
Despite your busy schedule, you were a gracious host the other day when I visited you while leading the Waseda Kendo Club. I am also deeply grateful for the memento you gave us.
Later, we traveled around the various regions of Hokkaido and returned to Tokyo without incident on the 23rd.
If I may be so bold, I expect that, under your guidance, the kendo world there will just get brighter and flourish even more.
Please provide your guidance regarding the future of Mr. Ichikawa.
Please take care of yourself at this time of the year. This is just to say thank you.

Signed: Saimura Gorō

P.S. I am sending you the journal of the Greater Youth Association, which the four of us organized. Please take a look.

"One Sword" – Saimura Gorō

On the occasion of the founding the Reigakudō, Sasamori Junzō, the 16th Sōke of Ono-ha Ittō-ryū, requested that Saimura Gorō, his close friend, brush these characters.

It still hangs there to this day as the soul of the dojo.

They say that Saimura Gorō had a deep knowledge of both Chinese and Japanese classical texts, and when he taught kendo, his instruction was infused with the core tenets of Eastern philosophy.

Swords of the of Ono-ha Ittō-ryū Patriarch's Repository

The Japanese sword evolved from having a straight blade to a curved one due to changes in the battlefield environment that occurred during the eras following the Heian Period.

In the Nanboku Period, the length of sword blades increased from three to five shaku as sword-making technology developed. However, during the peaceful times of the Edo Period, the bakufu issued an edict in 1882 that mandated that the length of the long sword would be two shaku, eight sun and that the length of the short sword would be one shaku, eight sun.

In Ono-ha Ittō-ryū, paired practice with wooden training swords is the norm, and since our school is known for having both the technique of kiriotoshi and also many stabbing techniques, swords with a shallow curvature (sori) are often used. Moreover, it is believed that the sword dimensions used in Ittō-ryū were established by the official kenjutsu instructors of the bakufu.

Many swords of the Ono-ha Ittō-ryū Patriarch's Repository were lost or misplaced when they passed between owner and successor during periods of transition. Yet, those that have remained in the collection are used in intense training even today. The koshirae and tsukas were made in the Edo Period, but the blades have been reforged many times up until the early Shōwa Period. We would like you to appreciate these swords not as works of art but to see the wear and tear on the sides of their blades that comes from our routine practice and recognize them as functional swords we use in actual training.

- **Ittō-ryū standard length sword: 2 shaku, 3 sun, 6 bu, 5 rin (71.4 cm)**
January, 1956; forged by Oshū Tsugaru resident, Kunitoshi; Reigakudō (NPO) Collection

The House of Nigara provided official swordsmiths to the Tsugaru Domain (present day Hirosaki City, Aomori Prefecture) for generations, and Nigara Kunitoshi was the 5th generation swordsmith of his family to work for them. He studied under Horii Toshihide and Kurihara Hikosaburō and was highly acclaimed for his style, which was reminiscent of the Rai school. He received the Minister of the Army Award at the Army Military Sword Exhibition in 1943. After the war, he suspended his smithing operations for a time but resumed forging swords in 1946 under the orders of the Commander, Eighth U.S. Army, when he became the U.S. Army's sword inspector. In 1954, he won the Prize for Excellence at the Society for Preservation of Japanese Art Swords. Following that, he won awards at numerous sword exhibitions.

In 1963, he was the first in Hirosaki City to be designated as a Living National Treasure, and in 1981, was designated as a Mukansa (a swordsmith regarded as so skilled that his work need not be appraised within the Society for Preservation of Japanese Art Swords).

The sword shown was made at the request of the 16th Sōke, Sasamori Junzō, who hailed from his same hometown. The hamon is suguha, there are small nie near the habuchi, there are bold nioi-guchi, the blade shines brilliantly, and the jigane is koitame hada, which, along with the jiba, is splendid and in good condition. Of all of Nigara Kunitoshi's works, this sword is regarded as the pinnacle of his achievement.

Ono-ha Ittō-ryū Patriarch's Repository: Kaneie Short Sword (Muromachi Period)

The Kamewari-tō sword, Itō Ittōsai's favorite, is famous among the swords of Ittō-ryū, but the reality is that its whereabout have been lost to history as it was passed from one headmaster to the next. Within the current sword collection that has been bequeathed, a Muromachi Period short sword made by Kaneie was discovered. Itō Ittōsai was alive during the Muromachi Period.

The Kaneie were swordsmiths whose work spanned from the Oei era (1394-1428) to the Kanei era (1624-1644). The 1st generation swordsmith is said to have been a student of Seki Zenjo Kaneyoshi. For successive generations, they signed swords as either Kaneie or Nōshū Seki-ju Kaneie. It is thought that they relocated to Echizen around the time of the 7th generation headmaster, after which they started to inscribed their names as Echizen Sonobe-ju Kaneie.

It is unquestionable that there are many discrepancies in the recorded history of Itō Ittōsai's life, and there are different theories about his place of birth, such as whether it was Izu-Itō or Izu-Ōshima, but according to Yamada Jirōkichi, the transmission documents of Kotōda Ittō-ryū assert that his place of birth was Katada, Ōmi. The founder of Kotōda Ittō-ryū, Kotōda Toshinao, became a direct student of Ittōsai after losing to him in a duel. These transmission documents are contemporaneous accounts of an old form of Ittō-ryū, so they are considered highly credible.

Kotōda called what he did Ittō-ryū throughout his entire life. It was his successors who started to call it Kotōda Ittō-ryū.

Also, though the ages have passed, according to a book handed down from the Edo Period called, *Ehon Eiyu Bidan*, published in 1881, Ittōsai was born in either Kanazawa of Kaga Province or Tsuruga of Echizen Province, and was the sword master of Otani Yoshitsugu, the lord of Tsuruga Castle. Otani Yoshitsugu, who was originally from Ōmi and later became the Lord of the Tusurga Domain of Echizen, may very well have hosted the renowned Itō Ittōsai since they both came from the same region. No records of his retainers remain, however, so this is nothing more than speculation.

Since it was appraised to have been forged during the Muromachi Period, the possibility that this Kaneie short sword of Ono-ha Ittō-ryū' s repository was passed from the Toda-ryū of Echizen to Itō Ittōsai, and then further handed down generation after generation starting with Ono Jirōemon Tadaaki, cannot be ruled out.

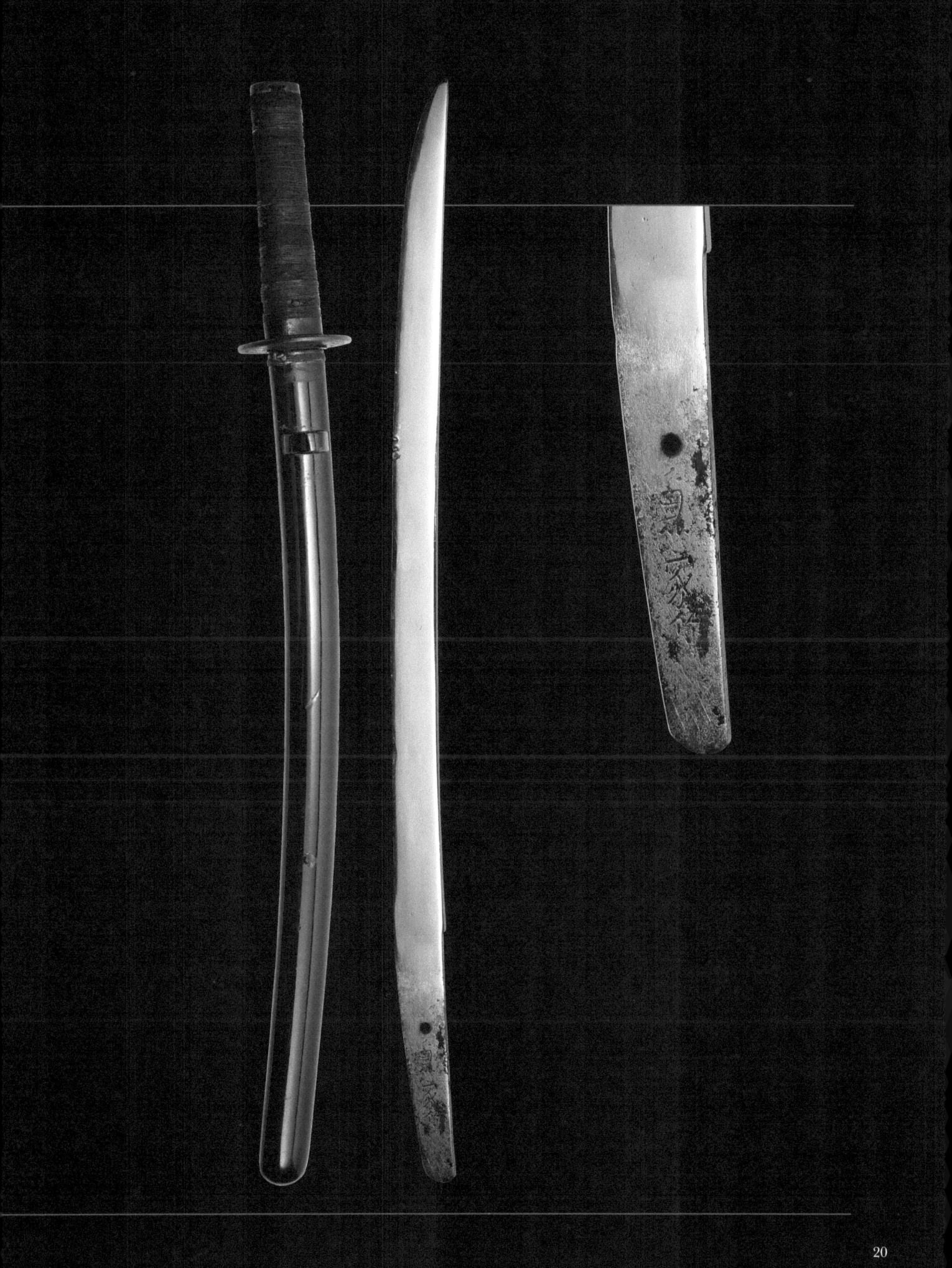

The Swords and Techniques of Ittō-ryū

The school of Ittō-ryū that maintains this sword repository is famous for having many techniques. Techniques such as kiriotoshi, mukaizuki, uki, suriage, and hari, are its foundation. It is also known for having many short sword techniques.

Among the swords that are currently stored within Ittō-ryū' s repository are two short swords from the Muromachi Period, many types of koshirae thought to have been from the end of the Bakumatsu era, and many blades that were forged after the beginning of the Shōwa Period. In fact, we still use them in our practice. In contrast to modern kendo, our approach to practicing kata is closer to actual combat. As a result, while the sword blades may look poorly maintained, they have simply withstood years of rough use.

Photo on the left
• Short sword koshirae with black lacquered scabbard and hard wood tsuka

  (Unsigned, Muromachi Period, koshirae from the Edo Period)

The tsuka resembles the Ukyo style. The koshirae is simple but well-balanced.

Photo on the right
• Short sword koshirae with black lacquered scabbard

  (Unsigned, Taishō and early Shōwa Periods)

Unlike the standard tsukamaki, this one is specially wrapped to make it sturdier and to keep it from coming unraveled during practice.

The tsuba is small and similar in shape to a Satsuma tsuba.

Uchigatana Koshirae With Black Lacquered Scabbard (Unsigned, Taishō to early Shōwa Periods)

Standard sized katana (koshirae for training use) mainly used during Tachiai Battō practice.

Unique characteristics of the sword include a tsuka that is wrapped in the jabara ito maki style and where the menuki is a little higher than on a standard uchigatana.

This is thought to prevent slippage and to increase power when doing kiriotoshi. The fuchigashira are the metal fittings of the fuchigane, which is connected to and touches the tsuba, and the metal fitting of the tsukagashira, which it matches on the opposite end. Both are plain red copper and made for practical use. The sword is unsigned but is thought to have been made sometime between the Taishō and early Shōwa Periods.

Uchigatana Koshirae with Black Lacquered Scabbard (Unsigned, Taishō to early Shōwa Periods)

Unlike a normal tsukamaki, this one is wound in a special way that makes it quite sturdy and difficult to come undone during practice. The sword is well made with little curvature of the blade and an elegant design.

「切落し」
Kiriotoshi

The way this cut is done is nearly the same as in modern kendo, but when performing it using a shinai without a shinogi, it turns into a technique close to uchiotoshi men, where both parties cut each other's forehead at the same time.

When doing this with a steel sword, rather than doing an uchiotoshi men, your cut slides down the side of your opponent's blade on its omote shinogi. Though the swords we use for practice have strong blades with blunted edges, the middle part of their blades along the omote shinogi is heavily worn due to practicing kiriotoshi like this.

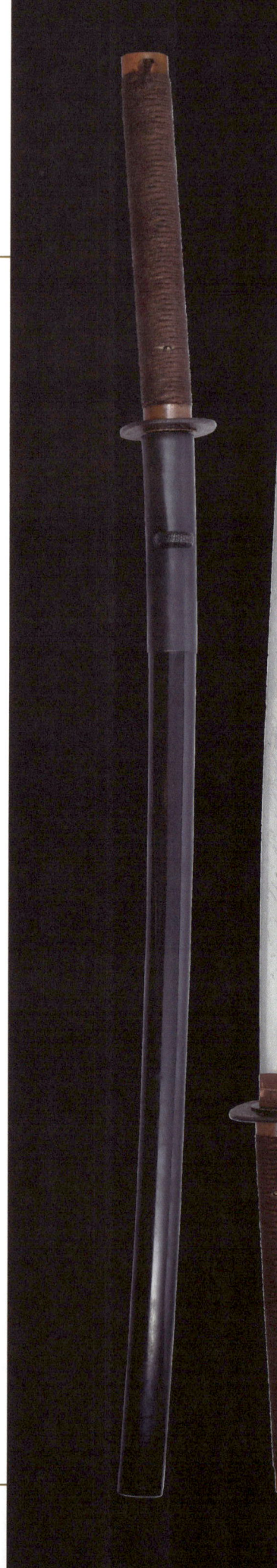

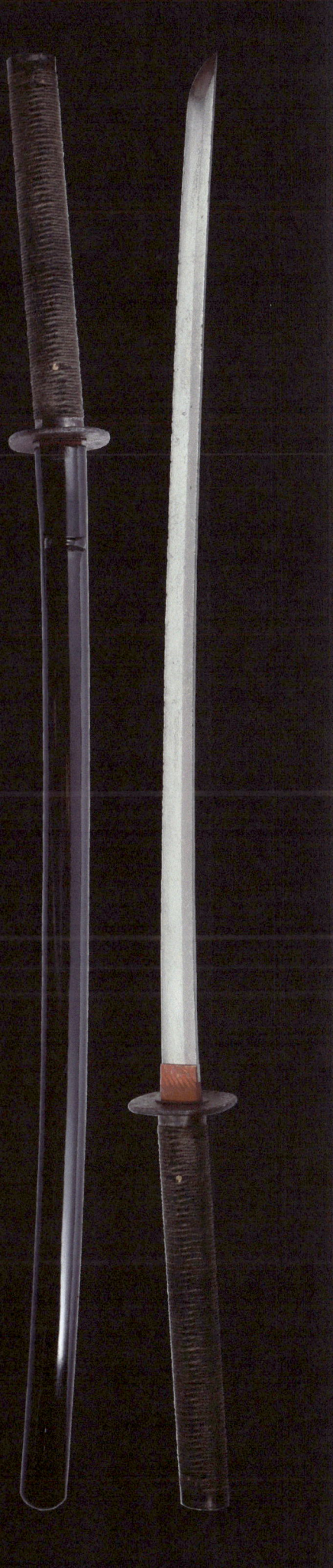

The tsukamaki is specially wound to make it sturdier and to keep it from unravelling during practice. It is well made with an elegant design.

「乗突」
Norizuki

This is a thrusting technique where you stab your opponent by turning your sword counter-clockwise so you can use the back of your blade against the side of his sword. The way the opponent's sword is wedged aside is the same as when doing kiriotoshi. The swords of Ittō-ryū are made with shallow sori, making them more suitable for doing stabbing techniques. On the uchigatana that is shown, you can see the many scuff marks on its omote shinogi between the tip and the middle of its blade due to this technique.

「浮木」
Uki

In this technique, you position your sword above your opponent's as if it were a log floating on water, and then use the power of your opponent against him. This is the basis of an attack where, when your opponent pushes down on your sword tip with all of his might, you wield your sword softly, borrowing his power to attack along his centerline. The deepest secrets of Ittō-ryū call for things to be done softly.

Uchigatana Koshirae with Black Lacquered Scabbard (Blade Unsigned, Taishō to Early Shōwa Periods)

「摺上」
Suriage

In Ittō-ryū, we don't just do a suriage against our opponent's blade from gedan, we also do a suriage from omote (as in Hoshatō) and ura (as in Ryūbigaeshi). We have many techniques that can be done in modern kendo.

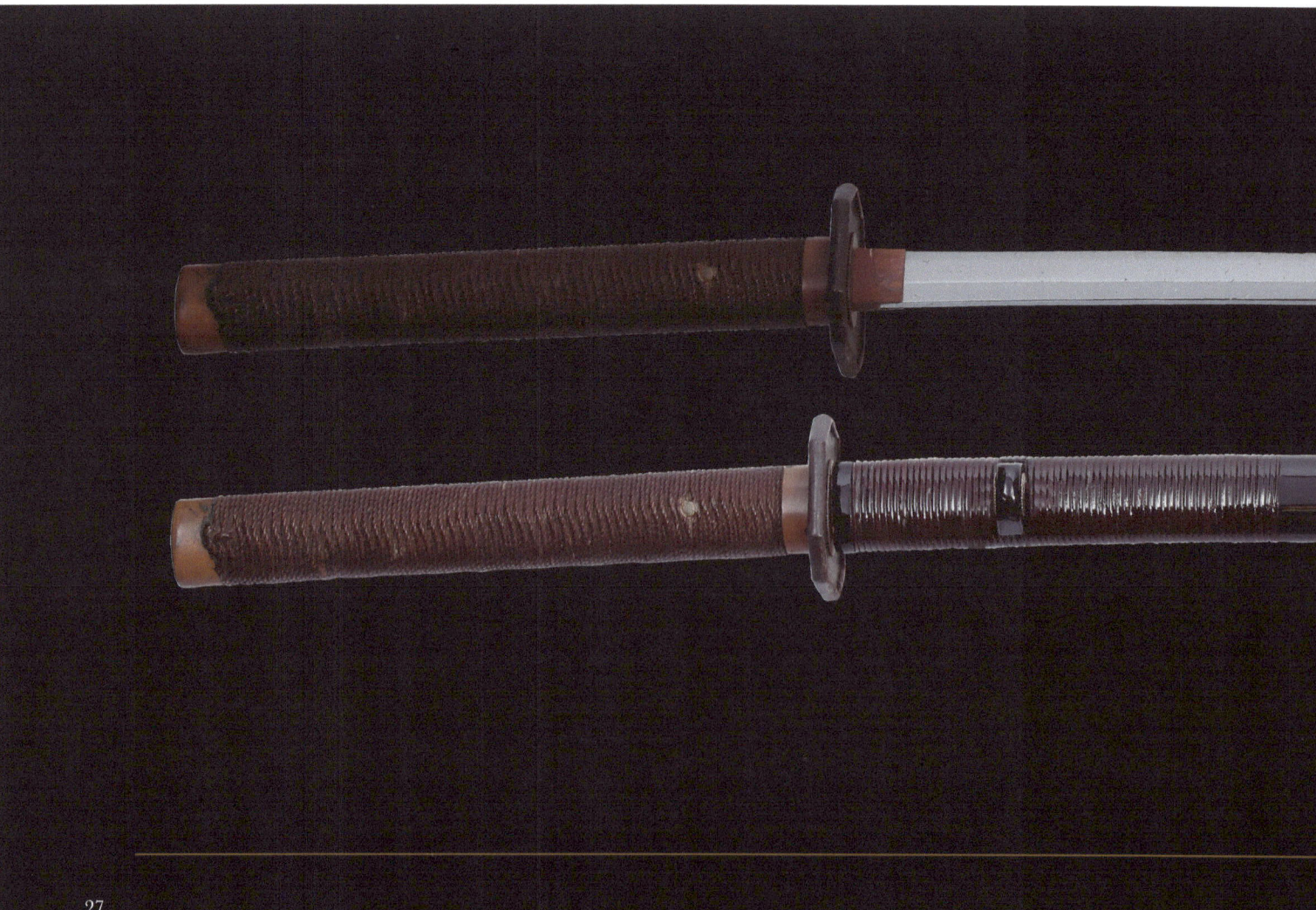

「張り」
Hari

Make a small, quick snapping motion with the wrists to slap your opponent's sword aside. Try to make this a smaller, sharper movement than a harai.

Shin Musō Hayashizaki-ryū Iai Odachi
— Three Shaku, Three Sun (100.8 cm)

Inscription: Shōwa 49 [1974], August, an auspicious day. Jointly produced by Yoshinobu and Kunitoshi, residents of Oshū Mutsu Province.
Tribute: presented to Sasamori Junzō, Hanshi of Ittō-ryū and Shin Musō Hayashizaki-ryū.
Jointly produced by the 4th master swordsmith Nigara Yoshinobu and the 5th master swordsmith Nigara Kunitoshi. An extremely high level of skill is required to make such a large blade. This habiki-style sword is used in training.
Reigakudō (NPO) Collection

Shin Musō Hayashizaki-ryū Iai was created through the ingenuity of Hayashizaki Jinsuke Shigenobu, the founder of the school who is also considered the creator of iaido.

"Shigenobu thought that when the two swords worn on one's side are drawn from their sheaths, they must always follow certain principles to ensure victory. He came to this realization after spending 100 days in spiritual reflection at the Tateoka Hayashizaki Myōjin Shrine. One day he was greatly enlightened after having seen a strange "manjiken" technique in a dream. Based on this revelation, he invented the art of sword drawing as a foundational method of sword fighting and called it Shin Musō Hayashizaki-ryū. The techniques of this school consist of: Mukōmi, Migimi, Hidarimi, Tonomono, Tonomono Yurushi, Nihōzume, Goka-no-Tachi, Hakka-no-Tachi, among others. During practice, it is a strange sight to see a sword that is 3 shaku, 3 sun long defeat a sword that is only 9 sun, 5 bu long."

This discipline is considered the origin of many iaido schools. A unique characteristic of it is that, during practice, the one performing the techniques always holds a sword that is 3 shaku, 3 sun (100.8 cm) long and uses it to cut down an opponent holding a short sword that is only 9 sun, 5 bu long just as he is about to attack with it.

Kumitachi

Itō Ittōsai developed unbeatable techniques for use with the long, medium, and short swords and came up with a way of preserving them within the curriculum of Ittō-ryū, passing on everything to his disciple, Ono Jirōemon Tadaaki. The overall curriculum, or kumitachi, of Ittō-ryū was finalized by Tadaaki, along with his descendants, Tadatsune and Tadao, and further refined by sword masters who came later. There are as many as 170 moves in the kumitachi, comprising: fifty moves of the Odachi kata plus ten supplemental moves, nine moves of the Kodachi kata, eight moves of the Aikodachi kata, one move of the Sanjū Nagaodachi kata, eleven moves of the Habiki kata, and the ten moves of the Hoshatō kata, which could actually be unlimited. In addition to these, there are the five moves of the Goten kata, twelve moves of the Hakiriai kata, nine moves of the Kuka-no-Tachi kata, eleven moves of the Taryūgachi-no-Tachi kata, seventeen moves of the Tsumeza Battō kata, and five moves of the Tachiai Battō kata. When you add to these the forty moves of the secret Kiyome-no-Tachi kata and the seven moves of the Gunshin Mihai-no-Shiki Tachi kata, the tally comes to over 170 moves.

When we practice the kumitachi, which is the foundation of Ittō-ryū, we use a special wooden practice sword that is thick and round but has a shallow sori.

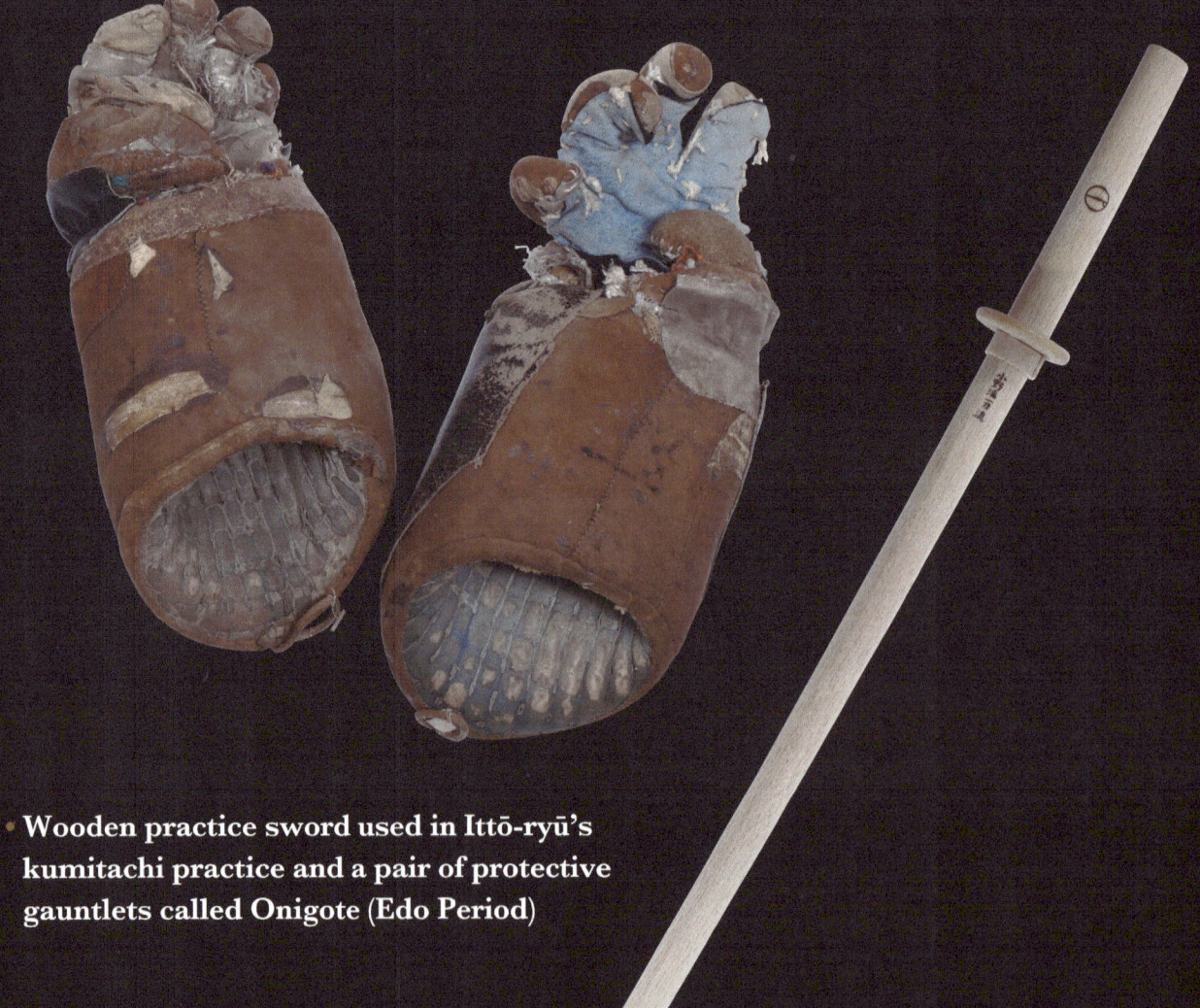

• Wooden practice sword used in Ittō-ryū's kumitachi practice and a pair of protective gauntlets called Onigote (Edo Period)

Ipponme (Hitotsugachi)

Ipponme is the very first move of Ittō-ryū and also its most secret, since it is where you learn how to do the technique of kiriotoshi. From time immemorial, Ittō-ryū has considered kiriotoshi to be such a fierce, strong, and righteous technique to ensure victory that we say, "Kiriotoshi is the beginning and kiriotoshi is the end." Because kiriotoshi appears so often throughout the curriculum, one learns it through the first move of Ipponme, where it must be fully developed, absorbed, and mastered. Kiriotoshi is not done by first striking down the opponent's sword and then cutting once again in two separate beats. It's done by observing your opponent's blade as it comes slashing down toward you while moving forward under it and beating him in a single beat without ever backing up. In other words, through the singular technique of cutting into your opponent, his sword is cut down to the side, which protects you. At that same instant, the momentum of your cut carries your sword forward to cut him in two. In short, one movement does the work of two. The act of cutting correctly forces your opponent's sword to the side, and the same instant his sword is diverted, your sword cuts straight into him. Because one move serves two functions, you always win.

If after his sword was diverted you were to attack again, or if you were to block a cut and then attack again, you would be using one to respond to one or two to respond to two, and the fight could go either way. It goes without saying that if you were to use two to respond to one, you would certainly lose.

Ittō-ryū's kiriotoshi teaches you how to use one to function as two. So, how do you win by cutting away your opponent's blade while you are both cutting toward each other at the same time? The secret to this is to first do kiriotoshi to your own mind.

1. Cutting down the opponent's blade with kiriotoshi.

2. Stabbing the opponent in the throat in the same motion.

3. The opponent's arm is cut just as he raises his sword to cut again.

Nihonme
(Mukaizuki • Norizuki • Tsukikaeshi)

The most terrifying feature of Ittō-ryū has long been considered its stabbing techniques, and the thrust taught through the second move of Nihonme is a useful technique to demonstrate what this really means. This move has several names, such as Mukaizuki and Tsukikaeshi, but the name that most accurately represents its true meaning is Norizuki. When your opponent's stab comes toward you, you must never retreat with the idea of evading or running away. If you were to do so, you would certainly be stabbed. If, instead, you were to judge the right time to thrust and then move forward while maintaining the proper kamae, your sword would ride over your opponent's, pushing his sword tip aside where it would "die," while he would end up impaled on the tip of your sword. Before you can do this technique, however, you first have to foster a keen eye and gritty determination. The resolve to act in the face of certain death leads to certain victory. Once your sword has stabbed your opponent and is positioned over his, even though you press down from above with both hands, don't try to put your weight onto your entire blade to hold his sword down. You should transfer your body weight to the area on your sword between your monouchi and sword tip, but how do you do that? The trick is to press down firmly with your right hand that is gripping your tsuka near the tsuba while lifting up your tsukagashira with your left hand. At the same time, grip the floor with the toes of your right foot and lift your heel slightly, while letting the heel of your left foot float upward to a certain extent. The weight of your body will transfer to your sword, which will then transfer your weight to your opponent's sword, holding it down. But you will not win by just holding down your opponent's sword. Your opponent will not like being controlled like this and will struggle to throw off the sword pressing down on him. As soon as his sword starts to go up to push yours away, suddenly pull yours away and an opening in his defenses will appear. Take advantage of that opening and strike. You have to learn how to release your opponent's sword after holding it down.

The genius of Ittō-ryū's Mukaizuki is on display when those who unjustly attack are impaled on the tip of your sword while you are in the seigan stance.

1. Doing Norizuki against Uchikata's thrust.

2. Riding the sword over Uchikata's and pressing it down.

3. Cutting Uchikata in two by striking his right jōdan just after letting his hands spring upward.

Tachiai Battō

Both Uchikata and Shikata use an odachi. When performing Tachia Battō, Uchikata initiates the attack, while Shikata subdues his attacker after defending himself. So, Shikata always draws his sword in response to Uchikata's actions, invariably winning. Shikata never draws his sword first to initiate an attack. When Shikata draws his sword in response to an attack, it is not merely to stop Uchikata's cut. Aside from protecting himself, Shikata's primary aim is to cut down his opponent the instant he pulls out his sword. The reason the swords form of a cross when they meet after they are drawn is because both sides are equally matched. Even though you may have the edge in terms of distance, fighting spirit, opportunity, or skill, drawing your sword doesn't necessarily guarantee victory. Only after you have an unwavering resolve to face death, driven by a burning passion for justice, can you justify drawing your sword. Therefore, you must diligently practice the skills of sword drawing and cutting while consistently nurturing your fighting spirit and strong moral convictions.

1. Ipponme Yūhi. Just as Uchikata draws his sword to do a reverse ura-kesagiri cut, Shikata does the same and blocks Uchikata's sword.

3. Ipponme Yūhi. Shikata just barely evades the thrust to his stomach, cutting down Uchikata with an omote kesagiri.

2. Ipponme Yūhi. From a gyaku hongaku stance, Uchikata thrusts towards Shikata's stomach.

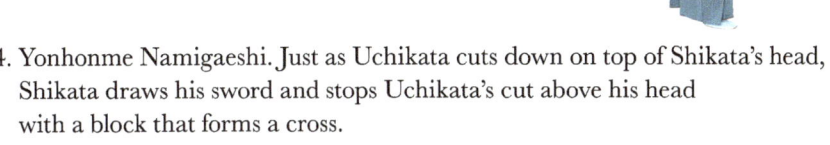

4. Yonhonme Namigaeshi. Just as Uchikata cuts down on top of Shikata's head, Shikata draws his sword and stops Uchikata's cut above his head with a block that forms a cross.

Tsumeza Battō

In Ittō-ryū, there are numerous short sword techniques that are derived from Chujō-ryū (Toda-ryū). The Tsumeza Battō kata contains techniques that involve getting up from a seated position while drawing your sword to defend against an attack. The corpus of our school contains thirty of these secret sword-drawing techniques.

During the Bakumatsu era, Sasaki Tadasaburō, said to be the best in Japan with a short sword, was a kenjutsu instructor at the Kōbusho, the shogunate's military academy. Commander of the Kyoto Mimawarigumi, he hailed from both Aizu and the Shintōseibu-ryū school of swordsmanship, but had been taught the five main schools of Aizu (Ittō-ryū Mizuguchi-ha, Ankō-ryū, Taishi-ryū, Shinten-ryū, and Shintōseibu-ryū) at the Aizu Domain's elite educational institute, the Nisshinkan. Because high ranking retainers of Aizu primarily learned Ittō-ryū Mizoguchi-ha, he is thought to have been influenced by Ittō-ryū.

Likewise, another master of the short sword, Katsura Hayanosuke (Nishioka Zeshin-ryū), is suspected to have been involved in the assassination of Sakamoto Ryōma while serving as a member of the Kyoto Mimawarigumi. The founder of the Nishioka Zeshin-ryū school, Nishioka Zeshin, is said to have been well versed in the subtleties of Ittō-ryū's short sword techniques. Tsumeza Battō, as the name suggests, are sword-drawing techniques used in cramped quarters. Techniques of cutting the opponent's eyes and ways of using the sword that involve placing one hand on the blade to increase cutting power are conveyed throughout.

1. Just as Uchikata draws his sword and cuts toward the right side of his opponent's neck,
   Shikata draws his sword, holds it vertically, and defends himself by throwing his opponent's sword off toward the right.

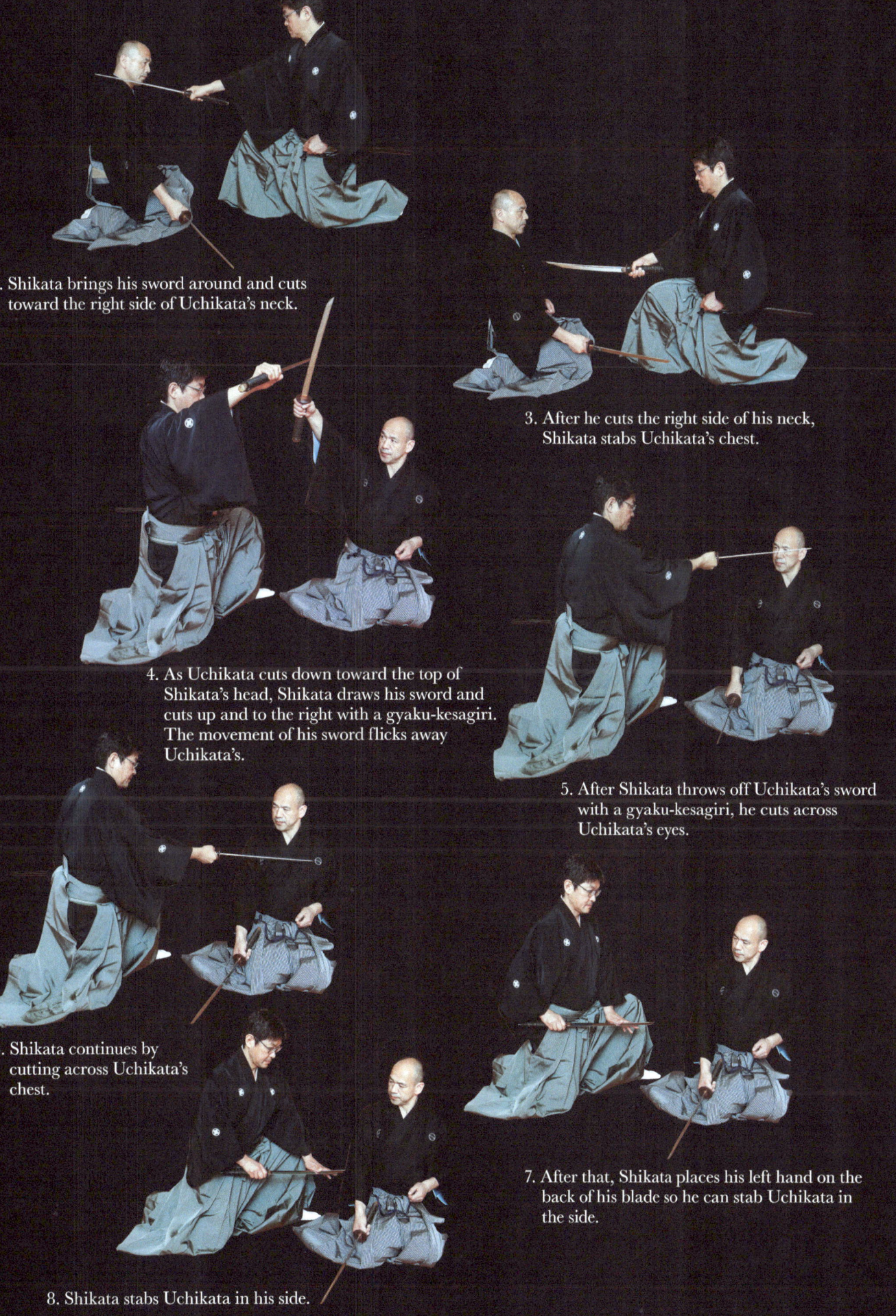

2. Shikata brings his sword around and cuts toward the right side of Uchikata's neck.

3. After he cuts the right side of his neck, Shikata stabs Uchikata's chest.

4. As Uchikata cuts down toward the top of Shikata's head, Shikata draws his sword and cuts up and to the right with a gyaku-kesagiri. The movement of his sword flicks away Uchikata's.

5. After Shikata throws off Uchikata's sword with a gyaku-kesagiri, he cuts across Uchikata's eyes.

6. Shikata continues by cutting across Uchikata's chest.

7. After that, Shikata places his left hand on the back of his blade so he can stab Uchikata in the side.

8. Shikata stabs Uchikata in his side.

Shin Musō Hayashizaki-ryū Iai

Sōke of Shin Musō Hayashizaki-ryū Iai
Ishizaki Toru

After this school was founded by Hayashizaki Jinsuke Shigenobu, it was passed on to Tamiya Heibei Akitsune, who mastered its secrets. The next who inherited it was Nagano Murakusai Kinro. Henceforth, it was passed on to Ichinomiya Sadayu Akinobu, Tani Kozaemon Norimasa, and Tsunei Kihei Naonori. Naonori passed the school to Lord Tsugaru Nobumasa, who embraced it and mastered its secrets himself. Asari Ihei Tadayoshi succeeded Naonori. After that, the school was passed on by the warriors of the Tsugaru Domain. Those who maintained the mainline within Tsugaru were: Tanaka Sōemon Harunori, Kon Mokuzaemon Hiromitsu, Kon Hachirōemon Hirotoshi, Kon Hachirōji Kanshu, Chiba Sōsōan, and Chiba Kennosuke. Around the time when Sasamori Junzō was the Chancelor of the Too-Gijuku, he invited Chiba Kennosuke to teach at his school and learned the art from him, eventually inheriting it. Sasamori Takemi learned the school from his father and succeeded him as its sole heir in 1956. After studying under Sasamori Takemi, Ishizaki Toru was designated by him to be the sole heir of the tradition. He became the current Sōke of Shin Musō Hayashizaki-ryū on October 1, 2016.

1. Mukōmi Ipponme Ottate: Shikata draws the sword horizontally and cuts.

• Shin Musō Hayashizaki-ryū Lineage

Hayashizaki Jinsuke Shigenobu − Tamiya Heibei Akitsune − Nagano Murakusai Kinro − Ichinomiya Sadayu Akinobu − Tani Kozaemon Norimasa − Tsunei Kihei Naonori − Asari Ihei Tadayoshi − Tanaka Sōemon Harunori − Kon Mokuzaemon Hiromitsu − Kon Hachirōemon Hirotoshi − Kon Hachirōji Kanshu − Chiba Sōsōan − Chiba Kennosuke − Sasamori Junzō − Sasamori Takemi − Ishizaki Toru

2. Mukōmi Ipponme Ottate:
   Shikata takes up a tenyoko ichimonji kamae.

3. Mukōmi Ipponme Ottate:
   Shikata finishes off his opponent using a
   nikyu cut.

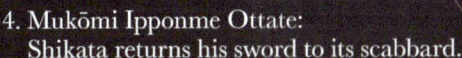

4. Mukōmi Ipponme Ottate:
   Shikata returns his sword to its scabbard.

5. Hidarimi Sanbonme Muchizume:
   Shikata draws his sword and cuts with an
   upward sweeping motion.

6. Hidarimi Sanbonme Muchizume:
   Shikata supports his sword with his left hand
   in a torii kamae and stabs Uchikata in his
   right ear.

7. Hidarimi Sanbonme Muchizume: Shikata
   supports his sword with his left hand and cuts
   across Uchikata's right arm as he draws his sword.

Tokugawa Iesato
— *Ryūro Muge* (alternatively, *Ruro Muge*)

Tokugawa Iesato (August 24, 1863 to June 5, 1940) was a Japanese politician and the 16th head of the Tokugawa dynasty. After the abolition of the feudal domain system, he became a member of the House of Peers, serving there for thirty years from 1903 to 1933, presiding as the 4th through 8th Presidents. He also took on such assignments as Ambassador Plenipotentiary to the Washington Naval Conferences on Disarmament, Chairman of the 1940 Tokyo Olympics Organizing Committee, 6th President of the Red Cross Society of Japan, Director of the Chinese Community Association, President of the Gakushuin Board of Councilors, Chairman of the Japan-America Society, Chairman of the 2,600th Anniversary Celebration of the Founding of Japan, among others. Sasamori Junzō was one of his acquaintances. Iesato studied at the Gakumonjo, the main educational institute of the Shizuoka Domain, and practiced Ittō-ryū swordsmanship under the tutelage of Asari Yoshiaki.

Ryūro Muge (or *Ruro Muge*)—characters penned by Tokugawa Iesato. In practicing the techniques of Ittō-ryū's kumitachi, one learns to release the stiffness and hesitation from their body, at which point their techniques become soft, large, and responsive.
When ice melts, it turns to water, and when rocks are crushed, they turn to sand, and both effortlessly take the shape of whatever container holds them. One should flow into the openings of their enemy's defenses without hesitation just like water. This is the meaning of *Ryūro Muge*.

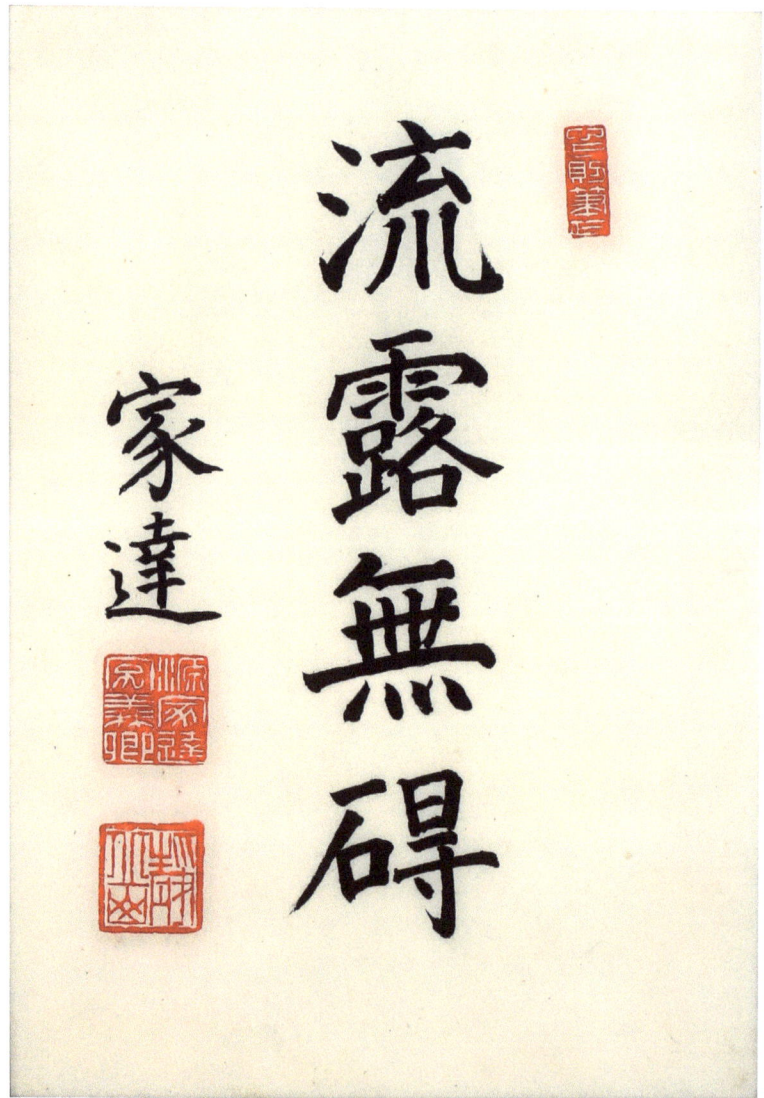

Portrait of Nakanishi Tsugumasa

Portrait and waka poem penned in his own hand by Nakanishi Chūbei Tsugumasa.

Nakanishi Tanesada, a master of Ono-ha Ittō-ryū, opened the Ittō-ryū Nakanishi Dojo on the east side of Neribei-koji Street in the Shitaya District of Edo. At the time, they practiced with wooden swords, but his successor, Nakanishi Tsugutake, upgraded the training by introducing protective gear between 1751 and 1763, after which they started to practice with bamboo swords. During the time of the 4th headmaster, Nakanishi Tsugumasa, the dojo was blessed with excellent students, to include Chiba Shūsaku, and gained the reputation as the best kenjutsu school in Edo. Nakanishi Tsugumasa awarded this hanging scroll together with a license of full initiation into the art to Sudō Hanbei, who later became a kenjutsu Shihan of the Hirosaki Domain. This work was passed to Sasamori Junzō from the Sudō family.

The waka poem that accompanies this portrait conveys the deepest secrets of kenjutsu.

[Translation]

Though I seek,
And though I seek again,
And though I seek,
I seek in vain,
The Way of kenjutsu

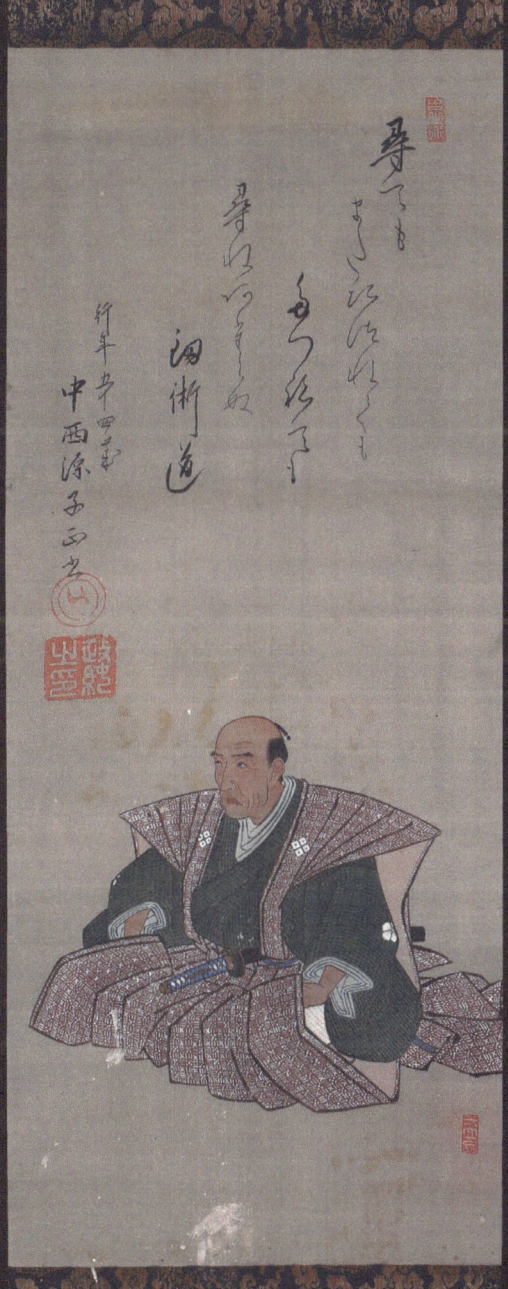

Correspondence from Tokugawa Iemitsu to Tsugaru Nobuyoshi

This is a letter in reply to a gift for a Boy's Day festival from Tsugaru Tosa-no-Kami to the House of Tokugawa. The seal of the third shogun, Tokugawa Iemitsu, is stamped on it, and the details referred to in the note are assumed to have been explained by Doi Oinokami Toshikatsu. The recipient, Tsugaru Tosa-no-Kami, was Tsugaru Nobuyoshi, the 3rd Lord of the Hirosaki Domain of Mutsu Province. The year this was composed was probably between 1631, when Tsugaru Nobuyoshi took over as lord of his domain, and 1638, when Doi Toshikatsu stepped down due to illness and assumed duties as a Grand Elder.

 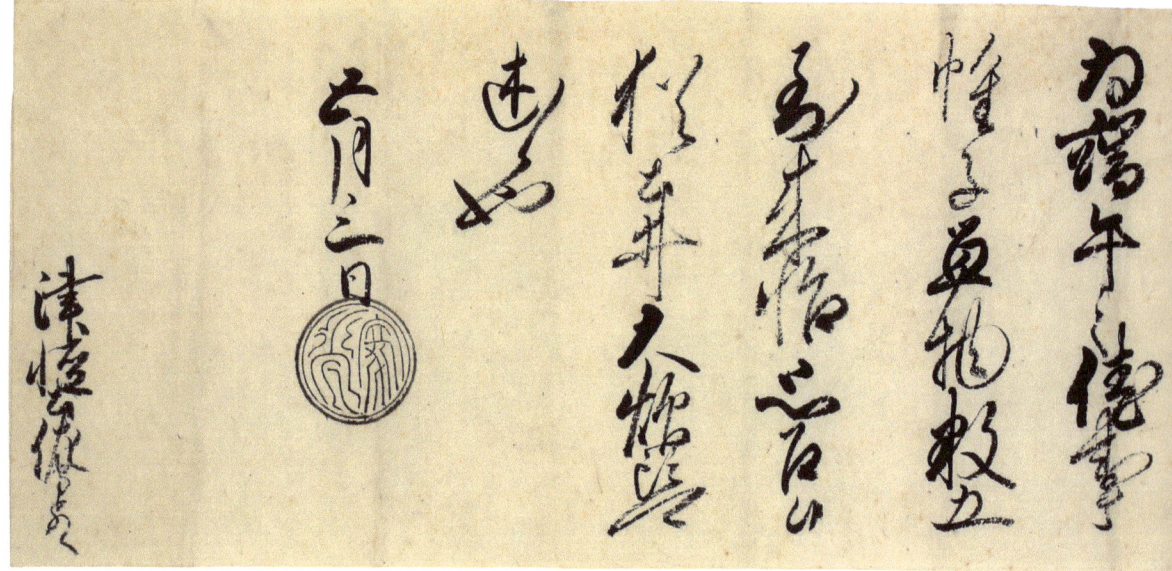

[Translation]

May 3

Tsugaru Tosa-no-kami:

I was glad to receive five rolls of material for the Boy's Day festival.
You can get more details from Doi Ōinokami.

Signed: Iemitsu

Correspondence from Sanjō Sanetomi to Yamaoka Tesshū

In response to the growing influence of Western civilization in Japan, the government began re-emphasizing traditional court rituals and ceremonies starting in 1882 as one measure to strengthen Japan's cultural and spiritual identity as a nation-state. You could say that this dissemination of traditional culture was planned out even within the world of kenjutsu, when, in 1883, the Yōyūkan and Saineikan were founded as part of the Peer's Club. These schools were established not just by Yamaoka Tesshū, but also by others who were connected to the Imperial Household, such as Sanjō Sanetomi, and other government institutions. With this historical context in mind, this letter is an invitation from Sanjō Sanetomi addressed to Yamaoka Tesshū asking him to come watch a gekken tournament that was planned to be held at the Yōyūkan. Yamaoka Tesshū was a student of Asari Yoshiaki, a kenjutsu master who had split off from Ono-ha Ittō-ryū. After the Meiji Restoration, Yamaoka founded the Mutō-ryū school of kenjutsu. This letter is part of Sasamori Junzō's collection, though it is unclear how he obtained it.

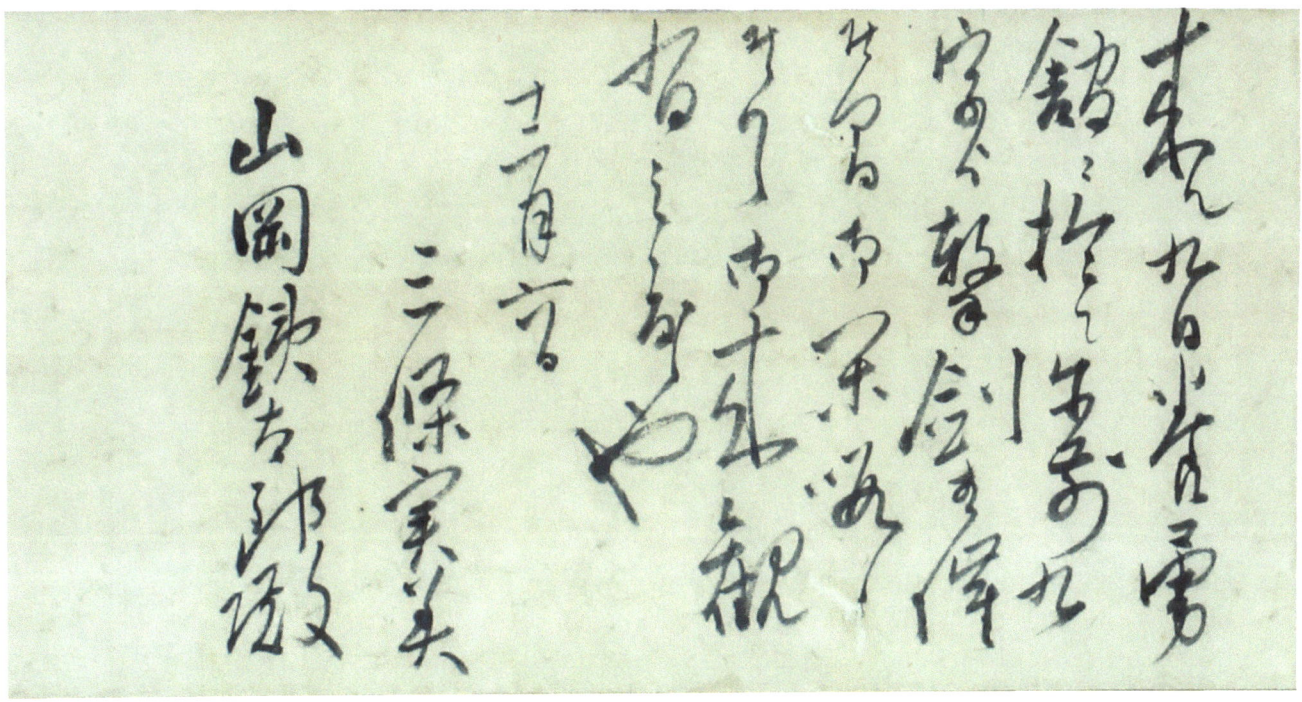

[Translation]

December 6

To: Yamaoka Tetsutarō

This coming 9th at the Yōyūkan, we will host a gekken event starting from nine o'clock in the morning. Please attend if you have time.

Signed: Sanjō Sanetomi

Ittō-ryū Kendo's Sixty-Six Moves for Certain Victory

After graduating from Waseda University, Sasamori Junzō, the 16th Sōke of Ono-ha Ittō-ryū, continued to practice the kumitachi with Takano Sasaburō, and there is no doubt that both student and teacher researched how the techniques of the kumitachi could invigorate kendo.

Nyūi Yoshihiro, a student of Takano Sasaburō, also taught his students thirty-six standardized techniques he compiled.

Moreover, Tokyo Metropolitan Police Honorary Shihan and Ono-ha Ittō-ryū menkyo-kaiden holder Ogawa Chūtarō Hanshi always used to say, "Learn techniques from Ono-ha Ittō-ryū!" Listed here are Ittō-ryū Kendo's Sixty-Six Moves for Certain Victory (twenty moves to strike men, twelve moves to strike the kote, seven moves to strike the dō, eighteen thrusting techniques, and nine combination techniques) that were identified by Sasamori Junzō.

Mastering the thrusting and combination techniques by actually trying them in the order listed is quite difficult even for practiced hands. Considerable wrist flexibility is required.

Section One: Men Waza

1. Kiriotoshi Men

When your opponent comes forward to cut your men, raise your sword even higher and step toward him, cutting down toward his men with kiriotoshi for the win.

2. Okorigashira Men

From aiseigan, strike your opponent's men just as he starts to come forward to do a technique.

3. Mukaejika Men

From aiseigan, go forward to meet your opponent as he comes forward to strike your men and immediately strike his.

4. Oikomi Men

Attack your opponent, pursue him as he escapes, step in and strike his men.

5. Nari Men

From aiseigan, when your opponent strikes toward your right kote, pull your sword back to your chest, raise it overhead, and then strike his men.

6. Nuki Men

From aiseigan, when your opponent tries to strike your right kote, avoid the cut by shifting to gedan, and then step forward and strike his men.

7. Hari Men

Use your sword to make a strong hari (bashing his sword to the side with a harai, nagashi etc.) and then step in and strike his men.

8. Makiotoshi Men

Use your sword to do a makiotoshi against his and then strike his men.

9. Ōjikaeshi Men

Move your sword to the left or right as your opponent cuts toward you, and then bring your hands around to strike his men.

10. Suriage Men

When your opponent cuts toward your men, use the side of your blade to do a suriage against his sword and then strike his men.

11. Harai Men

Do a harainoke against your opponent's sword that is cutting or stabbing. While you are doing that, step in and strike his men.

12. Kirikaeshi Men

When your opponent's sword comes in to strike your men, block it, bring your hands around, and then strike his men.

13. Katsugi Men

Bring your sword up toward your left shoulder in yō, and when your opponent defends his kote, spring forward and strike his men.

14. Chishō Men

As your opponent strikes toward your men, cut upward with your sword between his hands using chishō, and then immediately do a makiuchi to strike his men.

15. Kote Iro Kake Men

Act like you are going to strike your opponent's kote. Just as he lowers his sword to protect himself, jump in and strike his men.

16 Hidari Jōdan Men

Raise both hands in jōdan with the left foot and left hand forward. Strike your opponent's men when you see an opening in his seigan either with both hands or just the left. You can also strike his kote or dō.

17. Migi Jōdan Men

Raise both hands in jōdan with the right foot and right hand forward. Strike your opponent's men when you see an opening in his seigan either with both hands or just the right. You can also strike his kote or dō.

18. Katate Hazushi Men

When your opponent strikes toward your kote when you are in aiseigan, pull your right foot back, let go with your right hand, and extend your left hand up and to the left, and then strike toward the right side of your opponent's men. Or, let go with your left hand, extend your right hand toward the upper right and strike toward the left side of your opponent's men.

19. Sashi Men

When you push your opponent back as you attack, slide your right hand back to your tsukagashira while letting go with your left. Strike your opponent's men from outside range (tōma) by extending your right hand forward as you strike.

20. Hiki Men

From tsubazeriai, push your opponent back. When he backs away, raise your sword and strike his men.

Section Two: Kote Waza

1. Nami Kote

From aiseigan kamae, raise your sword tip above your opponent's and strike his right kote depending on the physical or psychological openings you see by. Use your sword tip to press your opponent's sword tip to the right or left and strike him when you see an opening.

2. Okorigashira Kote

Strike your opponent's right kote as soon as you see some indication he is about stab or strike.

3. Age Kote

As your opponent attempts to strike your men, strike his upraised right kote.

4. Iri Kote

Dip your sword tip from the right and circle it to the left under your opponent's tsuba as he stands in seigan. Insert your sword tip a little to the upper right and then strike his right kote.

5. Uchi Kote

If your opponent were to lower the tip of his sword in hira-seigan and position his sword to defend a strike from the outside, strike his uchi kote.

6. Katsugifuka Kote

Raise your sword toward your left shoulder in yō, and then get in close to strike your opponent's right kote from a side angle.

7. Hari Kote

When your opponent is standing in a kamae or when he is cutting or stabbing toward you, slap his sword to the right with a hari (this can also be a hajiki, harai, nagashi, uchiotoshi, suriage, etc.) and then strike his right kote.

8. Nuki Kote

When your opponent tries to strike your right kote, move to the left while lowering your sword to slip away from his cut. Then bring your sword around the right to strike his right kote.

9. Sasoi Kote

In aiseigan, act like you intend to strike your opponent's right kote, and when you clearly shift your right hand toward your left where he can see it, he will undoubtedly strike toward your kote. Wait for his strike and then sweep it away and strike his right kote.

10. Jōdan Kote

When, while in jōdan, you attack your opponent who is in seigan, he will defend by raising his sword tip at an angle. Spot the openings created by this move and strike his right kote.

11. Nobeshiki Kote

When your opponent is in jōdan, strike his forward-most kote from outside range (tōma) by extending your sword using nobeshiki.

12. Hiki Kote

From tsubazeriai, push your opponent toward your right front, and then while backing out to your left rear, strike his right kote.

Section Three: Dō Waza

1. Tobikomi Dō

Attack by stabbing upward, then jump forward and strike dō.

2. Nuki Dō (Hosha)

When your opponent strikes toward your men, step forward toward your right while dropping your body and striking the right side of his dō. Continue to slip out toward the right to get away. Or, when your opponent strikes toward your men, step forward toward your left while dropping your body and striking the left side of his dō. Continue to slip out to the left to get away.

3. Suriage Dō

When your opponent strikes toward your men, step forward toward your right while dropping your body and doing a suriage against his sword toward the upper left with the left side of your blade. Bring your hands around to strike the right side of his dō while continuing toward the right. Or, when your opponent strikes toward your men, step forward toward your left while dropping your body and do a suriage against his sword toward the upper right with the right side of your blade. Bring your hands around to strike the left side of his dō while continuing toward the left. You can also do a strong suriage to the left with the left side of your blade and then strike the left side of his dō while continuing toward the right.

4. Hari Dō

Do a hari, hajiki, nagashi etc. from either the left or right against your opponent's sword as it comes forward, then immediately step forward and strike the left or right side of his dō.

5. Jōdan Seme Dō

From jōdan, strike toward your opponent's men. Strike his dō as he raises his hands.

6. Kote Kake Dō

From seigan, attack toward your opponent's kote while he is in jōdan. Just as he moves to avoid the strike, step forward and strike his dō.

7. Temoto Hiki Dō

When pressing up against your opponent to the right or left in tsubazeriai, strike the left or right side of his dō as you back up.

Section Four: Tsuki Waza

1. Kiriotoshi Tsuki (Deba Kiriotoshi)

When your opponent tries to strike your men, go toward him and stab him with the deba variation of kiriotoshi.

2. Mukai Tsuki

When your opponent stabs you, go forward to meet it, turning the edge of your sword blade to the right and pressing down on top of his from above.

3. Nori Tsuki

From aiseigan, keep your left hand low and supported by your right hand while keeping the edge of your blade facing straight down. Firm up both hands and go forward with resolve as you ride your sword over your opponent's tsuba, stabbing him.

4. Omote Tsuki

From aiseigan, turn the edge of your sword blade to the right, and step forward while stabbing him from omote.

5. Ura Tsuki

From aiseigan, turn the edge of your sword blade to the left, and step forward while stabbing him from ura.

6. Otoshi Tsuki

From aiseigan, raise your sword tip slightly and then stab as you drop it.

7. Hari Tsuki

From aiseigan, use your sword to do a hari, hajiki, uchiotoshi, etc. against your opponent's sword when he is in a static kamae or while attacking, and then stab him.

8. Mawashi Tsuki

From aiseigan, circle your sword tip around your opponent's sword to the left and right to confuse him as to which way you plan to attack, and then stab him.

9. Nuki Tsuki

When your opponent moves back to avoid your attack, step in toward him and stab.

10. Nayashiire Tsuki

If your opponent shrinks back from your attack, step in even deeper and stab.

11. Rishō Tsuki

If, when your opponent comes forward to apply a technique, you thrust both hands forward with resolve, your opponent will simply impale himself on your sword tip.

12. Kote Iro Tsuki

Act as if you are trying to strike his right kote and then stab him just as he drops his sword tip to avoid the strike.

13. Kote Hazushi Tsuki

When your opponent strikes toward your right kote, let go with your right hand and make a one-handed stab with your left.

14. Kote Kake Tsuki

When your opponent goes to strike your men, hold his right kote down with your sword and then stab him.

15. Chishō Tsuki

As your opponent approaches to strike your men, raise your sword up between his hands using Chishō, turn your hands over, and then stab him.

16. Uki Tsuki

When your opponent approaches to strike your men and you do uchiotoshi, if he were to rotate your sword, stab him using the thrust from Uki.

17. Nidan Tsuki

In aiseigan, thrust forward from omote. When your opponent tries to avoid the stab by pushing it toward omote, bring your sword around to the other side, step forward, and then stab him from ura.

18. Sandan Tsuki

From aiseigan, when you quickly drop your sword tip to attack from ura, your opponent will defend to that side. When you attack from omote, your opponent will defend omote. At that point, bring both hands around to ura, step in, and stab. This is a technique that has to be done quickly and decisively.

Section Five: Combination Techniques

1. Kote Kake Men

From aiseigan, go to strike your opponent's kote, and just as he drops his sword tip to defend, strike his men.

2. Kote Men · Dō · Tsuki

From aiseigan, strike toward your opponent's kote and then continue on to strike his men. Or strike his dō and finish him off with a tsuki. Combination techniques are done quickly and with intensity.

3. Kote · Katate Tsuki · Dō

Strike the kote and stab with the left hand, and then regrip with the right hand and slash his dō.

4. Suriage Kote · Tsuki

When your opponent strikes toward your kote, do a suriage to the right and strike his right kote. As you strike, stab.

5. Hari Men · Dō

Slap your opponent's attack away with a hari-yaburi, strike his men, and then slash his dō.

6. Katate Kote · Men

From a right one-handed jōdan, strike your opponent's right kote when he is in seigan. Bring your sword around and strike his men.

7. Migi Katate Migi Men · Hidari Men

When your opponent strikes toward your right kote, let go with your left hand, do a suriage to your right with your right hand, and strike the left side of his men. Immediately switch the sword to your left hand and strike the right side of his men.

8. Hajiki Age Kote · Dō

When your opponent strikes toward your men, flick his sword away using a hajiki and then strike his right kote. Bring your hands around to strike the right side of his dō while going past him toward your right.

9. Suriage Migi Dō · Hidari Dō · Tsuki

Do a suriage against your opponent's sword that is cutting toward your men and then strike his right dō. Immediately bring your hands around and strike the left side of his dō and then stab with both hands.

*Ashigarami is short for tai-atari and kumiuchi-no-te.

Spirit of the Reigakudō

The spirit that fills and is shared within the Reigakudō was imparted to us by Sasamori Takemi, the 17th Sōke of Ono-ha Ittō-ryū, and is the spirit that has been passed down within the Hirosaki Domain from the House of Ono to the Houses of Tsugaru, Yamaga, and Sasamori.

Within the Hirosaki Domain, the Ono-ha Ittō-ryū that was handed down from Ono Jirōemon Tadakazu to the 5th Lord of Tsugaru, Nobumasa, was later conveyed to Yamaga Hachirōzaemon Takayoshi, and then from within the House of Yamaga, it was passed down to Sasamori Junzō from Yamaga Motojirō Takatomo.

The Yamaga family of the Hirosaki Domain were the descendants of the Confucian and military scholar Yamaga Sokō, and they disseminated Yamaga-ryū strategy and the spirit of bushido throughout the province. The bushido embraced throughout the domain, deeply influenced by Yamaga Sokō's dictum that, "The way of the sages is not one man's personal possession," promoted sage learning as something meant for all people, not just the warrior class.

Sasamori Junzō's father was a retainer of the Hirosaki Domain who served, among other duties, as the clan's Chief Master at Arms and was one of those heavily influenced by his homeland's spirit of bushido.

Sasamori Junzō shared his father's strong embrace of this philosophy and, before World War II, traveled by himself to the United States to study. Later, he was entrusted with the responsibility of rejuvenating the Too-Gijuku as its Chancelor, and after that, took up another post as the Chancelor of Aoyama Gakuin. After the war, he worked tirelessly to revitalize kendo and help Japan recover as a post-war Minister of Home Affairs.

His son Takemi saw his father's bushido spirit up close and was one of those inspired by it. Takemi was always kind, never once raising his voice in anger toward the students he was teaching. However, behind his eyes there was always a certain intensity.

Takemi used to explain that when he taught that an aggregate of sharp corners forms a circle, he was saying that true kindness required pointedness. At first glance, these appear to be ideas at odds with each other, but at their core, they are same, and we call the state in which many pointy corners overlap each other to form a completely filled out circle, *harmony*, or *enman*. *Pointedness* is critically reflecting on your own actions, and the ability to be empathetic toward others is not possible unless you take a hard look at yourself. If you don't do this, you will never be considerate toward others. There is no doubt that we strongly feel the spirit of Ittō-ryū's "forge yourself without fighting," through the "silent teachings," and we will continue to pass on these lessons.

The 17th Sōke, Sasamori Takemi, who was also a protestant minister, presented his views on the philosophical underpinnings shared by both Christianity and bushido after researching them both over the span of many years. The seppuku of the warrior is akin to martyrdom. Justice is akin to love. The profound significance of the phrase found in the *Hagakure, the way of the warrior is found in death,* argues that others live through the spirit of martyrdom found in self-sacrifice. What are the common denominators between the Christian faith and the spirit of bushido? Looking back on the history of Christianity after the Meiji Period, how will it survive into the future? Based on his personal experiences and testimony starting from when he was a child about the *Way* and *True Way of Living*, he deeply explored the philosophical contradictions between the two "Ways" that question life and death: the Christian who preaches peace and love and the samurai who was devoted to fighting. His book *Bushido and Christianity* (Shincho Shinsho) has been highly acclaimed as an introduction to Christianity that addresses these questions.

Sasamori Takemi

May 10, 1933 to August 15, 2017. Protestant minister, 17th Sōke of Ono-ha Ittō-ryū, and Executive Director of the Reigakudō. Born in Aomori Prefecture as the third son of Sasamori Junzō. Also inherited Chokugen-ryū Naginata-jutsu and Shin Musō Hayashizaki-ryū Iaijutsu. Earned an undergraduate degree from Waseda University and a graduate degree from the United States' Duke University Graduate School of Divinity. Operating out of the Komaba Eden Church, he performed duties as both a protestant minister and Executive Director of the Japan Kobudō Kyōkai. Author of *Bushido and Christianity; Kami e no michi; Kami kara no michi,* and others.